x64 汇编语言：从新手到 AVX 专家

[比] 乔·范·霍伊(Jo Van Hoey) 著
贾玉彬 王昱波 译

清华大学出版社

北京

北京市版权局著作权合同登记号图字：01-2021-5176

Beginning x64 Assembly Programming: From Novice to AVX Professional, First Edition by Jo Van Hoey

Copyright © Jo Van Hoey, 2019

This edition has been translated and published under licence from Apress Media, LLC, part of Springer Nature.

本书中文简体字版由 Apress 出版公司授权清华大学出版社出版。未经出版者书面许可，不得以任何方式复制或抄袭本书内容。

本书封面贴有清华大学出版社防伪标签，无标签者不得销售。
版权所有，侵权必究。举报：010-62782989，beiqinquan@tup.tsinghua.edu.cn。

图书在版编目(CIP)数据

  x64 汇编语言：从新手到 AVX 专家 /(比)乔·范·霍伊(Jo Van Hoey) 著；贾玉彬，王昱波译. —北京：清华大学出版社，2022.1（2023.9重印）
  书名原文：Beginning x64 Assembly Programming:From Novice to AVX Professional, First Edition
  ISBN 978-7-302-59546-5

  I. ①x… II. ①乔… ②贾… ③王… III. ①汇编语言—程序设计 IV. ①TP313

  中国版本图书馆 CIP 数据核字(2021)第 231296 号

责任编辑：王　军　韩宏志
装帧设计：孔祥峰
责任校对：成凤进
责任印制：刘海龙

出版发行：清华大学出版社
       网　　址：http://www.tup.com.cn, http://www.wqbook.com
       地　　址：北京清华大学学研大厦 A 座　　邮　　编：100084
       社 总 机：010-83470000　　邮　　购：010-62786544
       投稿与读者服务：010-62776969, c-service@tup.tsinghua.edu.cn
       质 量 反 馈：010-62772015, zhiliang@tup.tsinghua.edu.cn

印 装 者：三河市少明印务有限公司
经　　销：全国新华书店
开　　本：170mm×240mm　　印　张：23　　字　数：451 千字
版　　次：2022 年 1 月第 1 版　　印　次：2023 年 9 月第 3 次印刷
定　　价：88.00 元

———————————————————————————————————————

产品编号：091910-01

译者序

收到王军老师发来的英文电子版时，我的内心是忐忑的。汇编语言是一门低级语言，晦涩难懂，学习门槛高。目前市面上的汇编语言书籍大多是 32 位的，64 位汇编语言的书籍凤毛麟角，可供查证的资料十分有限。我于 2020 年 3 月开始正式创业；作为上海碳泽信息科技有限公司的创始人，我平时工作较忙，只能利用周末和晚上的时间进行翻译，翻译这样的书籍是很有挑战的。我尽量把书中的示例代码都亲自跑了一遍，以进行验证。本书的作者有非常丰富的从业经验，全书内容深入浅出，循序渐进，通俗易懂；本书的技术审稿人在英特尔公司工作了 26 年，对英特尔 CPU 的架构非常熟悉，对本书的质量进行了严格把关。

汇编语言是逆向工程的基础，扎实的汇编语言基础对于抵御恶意软件攻击非常有用，是保护 IT 基础设施的重要技能。在学习本书前，最好先学一门高级语言(如 C 语言)，这对于你学习和掌握本书的内容非常重要。本书很少涉及深奥的理论知识，将理论内容控制在最低范围。书中的代码都是完整的，便于读者进行测试和修改。

掌握汇编语言对于漏洞挖掘也非常重要，尤其是底层漏洞，如 Windows、Linux 操作系统的漏洞挖掘。另外，如果读者喜欢打 CTF 比赛，那么任何一个优秀战队都需要一名优秀的二进制队员；要想成为那个不可或缺的队员，非常有必要学好汇编语言。

目前国内的安全人才(尤其是二进制方面的人才)非常短缺。希望本书能帮助读者快速掌握 64 位汇编语言，迅速上手，为精通 64 位汇编语言打下坚实基础，为国家信息技术的发展和网络安全事业奉献力量。

最后感谢清华大学出版社，感谢王军等编辑付出的艰苦努力，感谢上海碳泽信息科技有限公司的所有同事和股东的支持，谢谢你们。

贾玉彬

2021 年 7 月于北京

作者简介

Jo Van Hoey 拥有 40 年的 IT 行业从业经验，包括各种职能部门、多家 IT 公司和各种计算平台。他最近从 IBM 大型机软件客户经理职位上退休。出于对 IT 安全的兴趣，Jo 长期深入研究汇编语言，因为汇编语言知识是保护 IT 基础设施抵御攻击和恶意软件的重要基础。

技术审稿人简介

Paul Cohen 在 x86 架构的早期(从 8086 开始)就加入了英特尔公司，在销售/市场/管理领域工作了 26 年后从英特尔退休。他目前与道格拉斯技术集团(Douglas Technology Group)合作，专注于代表英特尔和其他公司创作技术书籍。Paul 还与 YEA 学院合作，讲授一门帮助初中生和高中生蜕变为真正的、自信的企业家的课程。他是俄勒冈州比弗顿市的交通专员，也是多个非营利组织的董事会成员。

前　言

学习汇编语言可能会令人沮丧，不仅仅是因为它是一种"无情"的语言。在任何可能的情况下，计算机都会对你"发飙"。你可能只是在不知不觉中引入了一个隐藏的错误，该错误会在以后的程序中或执行时让你发狂。除此之外，学习曲线陡峭，语言晦涩难懂，英特尔官方文档铺天盖地，并且可用的开发工具各有其古怪之处。

在本书中，你将从简单的程序开始学习，直到高级矢量扩展(AVX)编程。到本书结尾，你将能编写和阅读汇编代码，将汇编语言与更高级的语言混合在一起使用，并对 AVX 有初步的了解。本书的目的是向你展示如何使用汇编语言指令。本书不是关于编程风格或代码性能优化的。在掌握汇编的基本知识之后，你可以继续学习如何优化代码。本书不应该是你学习编程的第一本书。如果你没有任何编程经验，请将本书搁置一段时间，先学习更高级语言(如 C)的一些编程基础知识。

可扫描封底二维码下载本书的所有源代码。本书使用的代码都尽可能地简单，这意味着没有图形用户界面、繁杂的程序或错误检查，因为添加所有这些特性会使我们的注意力偏离学习汇编语言的目的。

理论知识被严格限制在最低限度：关于二进制数的一点知识、逻辑运算符的简短介绍以及一些有限的线性代数。我们将远离浮点转换。如果你需要转换二进制或十六进制数，请找一个可以为你执行此类操作的网站。不要浪费时间执行手工计算。坚持目标：学习汇编。

汇编代码以完整的程序呈现，这样你就可以在计算机上测试它们、操作它们、更改它们、破坏它们。

还将展示可以使用哪些工具、如何使用它们以及这些工具的潜在问题。拥有正确工具对于克服陡峭学习曲线至关重要。有时会指出可能有用或提供更多详细信息的书籍、白皮书和网站。

我们无意为你提供所有汇编指令的综合课程。内容太多了，仅用一本书讲述是不可能的(请看英特尔手册的大小)。我们将介绍主要的指令，以便你对汇编语言有所了解。如果通读本书，你将获得学习汇编的必要知识，你还可以自行详细研究某些感兴趣的领域。读完本书后，你将可以进一步学习英特尔手册，并尝试理解其内容。

本书的大部分内容专门针对 Linux 上的汇编，因为它是学习汇编语言的最简单平台。最后，我们提供了一些章节以帮助你了解如何在 Windows 上进行汇编。 你将看到，一旦掌握了 Linux 汇编，进行 Windows 汇编会更容易。

有许多可与英特尔处理器一起使用的汇编器，例如 FASM、MASM、GAS、NASM 和 YASM。我们将在本书中使用 NASM，因为它是多平台的，在 Linux、Windows 和 macOS 上都可用，具有较大的用户群。但不必担心，一旦你熟悉一种汇编器，学习另一种汇编"方言"就很容易。

我们已经仔细编写并测试了本书使用的代码。但是，如果文本中有错别字或程序中有错误，我们将不承担任何责任。我们将错误归咎于两只猫，它们喜欢在我们打字的时候走过我们的键盘。

我们在本书中提出的想法和观点仅代表我们自己，并不代表 IBM 的立场、战略或观点。

开始之前

开始阅读本书之前你需要掌握以下基础知识：
- **安装和使用虚拟化软件(VMware、VirtualBox)**。如果你对虚拟化一无所知，可以从 https://www.virtualbox.org 下载免费的 Oracle VirtualBox 并进行安装。然后通过在其中安装 Ubuntu Desktop Linux 来学习如何使用虚拟化软件。可以通过虚拟化软件在你的主机上安装不同的操作系统。通过虚拟机进行实验时不会对你的宿主机系统造成损害。互联网上有大量资料介绍 VirtualBox 和其他虚拟化软件解决方案。
- 你需要具备 Linux 命令行界面(CLI)的基本知识。我们将在本书中一直使用命令行。如有必要，可以使用另一个 Linux 发行版，但要确保可以安装书中使用的工具(NASM、GCC、GDB、SASM 等)。以下是你需要掌握的基本知识：如何安装操作系统，如何安装其他软件，如何通过命令提示符启动终端，以及如何在命令行上创建、移动、复制和删除目录及文件。你还需要知道如何使用 tar、grep、find、ls、time 等命令以及如何启动和使用文本编辑器。不需要了解高级 Linux 知识；你只需要了解基本知识，就可以按照本书中的说明进行操作。如果你不了解 Linux，请尝试使用并适应它。互联网上有很多很好的、简短的入门教程(例如 https://www.guru99.com/unix-linux-tutorial.html)。你将看到在 Linux 机器上学习了汇编后，在另一种操作系统上学习汇编也比较简单。
- 你应该具有 **C 编程语言的一些基本知识**。我们将使用几个 C 函数来简化示例汇编代码。此外，我们还将展示如何与高级语言(例如 C)进行交互。如果你不懂 C 语言，但想充分享受这本书，可以参加一些免费的 C 语言入门课程，如 tutorialspoint.com。不必完成全部课程，只要看一下这门语言的几个程序就可以了。你可以稍后再来了解更多细节。

为什么要学习汇编语言

学习汇编语言有以下好处：
- 你将学习 CPU 和内存是如何工作的。
- 你将学习计算机是如何和操作系统一起工作的。
- 你将学习高级语言编译器如何生成机器语言，并且这些知识可以帮助你编写更高效的代码。
- 你将可以更好地分析程序中的错误。
- 当你最终让程序运行起来的时候，是非常有趣的。
- 你想调查恶意软件，但你只有机器代码，没有源代码。了解汇编语言后，你将能够分析恶意软件并采取必要的操作和预防措施。

英特尔手册

英特尔手册包含所有你想知道的关于英特尔处理器编程的知识。然而，这些知识对于初学者来说是难以接受的。当你继续阅读本书时，你将看到英特尔手册中的解释将逐渐对你更有意义。我们将经常提到这些海量知识。

要查看英特尔手册，可访问 https://software.intel.com/en-us/articles/intel-sdm。不要把它们打印出来，因为内容非常多！浏览一下这些手册，看看它们有多么全面、详细和正式。通过这些手册学习汇编是非常困难的。我们特别感兴趣的是第 2 卷，你将在其中找到有关汇编编程指令的详细说明。

你可以在这里找到有用的来源：https://www.felixcloutier.com/x86/index.html。这个网站提供了所有说明的列表以及如何使用它们的摘要。如果这里提供的信息不够，你可以随时使用英特尔手册或互联网搜索工具。

目 录

第 1 章 你的第一个程序 ………………… 1
  1.1 编辑、汇编、链接和运行(或调试) ……………………………… 2
  1.2 汇编程序的结构 …………………… 6
    1.2.1 .data 段 ………………………… 6
    1.2.2 .bss 段 ………………………… 7
    1.2.3 .txt 段 ………………………… 8
  1.3 小结 ………………………………… 10

第 2 章 二进制数、十六进制数和寄存器 ……………………………… 11
  2.1 二进制简短课程 …………………… 11
    2.1.1 整数 …………………………… 12
    2.1.2 浮点数 ………………………… 13
  2.2 寄存器简短课程 …………………… 13
    2.2.1 通用寄存器 …………………… 14
    2.2.2 指令指针寄存器(rip) ………… 15
    2.2.3 标志寄存器(Flag Register) ………………………… 15
    2.2.4 xmm 和 ymm 寄存器 ………… 16
  2.3 小结 ………………………………… 16

第 3 章 用调试器进行程序分析：GDB ……………………………… 17
  3.1 开始调试 …………………………… 17
  3.2 继续进步 …………………………… 22
  3.3 其他 GDB 命令 …………………… 24
  3.4 稍加改进的 hello, world 程序 ……………………………… 25
  3.5 小结 ………………………………… 27

第 4 章 你的下一个程序：Alive and Kicking …………………… 29
  4.1 alive 程序分析 ……………………… 30
  4.2 打印 ………………………………… 34
  4.3 小结 ………………………………… 37

第 5 章 汇编是基于逻辑的 …………… 39
  5.1 NOT ………………………………… 39
  5.2 OR …………………………………… 39
  5.3 XOR ………………………………… 40
  5.4 AND ………………………………… 41
  5.5 小结 ………………………………… 42

第 6 章 数据显示调试器 ……………… 43
  6.1 使用 DDD ………………………… 43
  6.2 小结 ………………………………… 46

第 7 章 跳转和循环 …………………… 47
  7.1 安装 SimpleASM ………………… 47
  7.2 使用 SASM ………………………… 47
  7.3 小结 ………………………………… 54

第 8 章　内存 ·············· 55
- 8.1　探索内存 ·············· 55
- 8.2　小结 ·············· 62

第 9 章　整数运算 ·············· 63
- 9.1　从整数算术开始 ·············· 63
- 9.2　分析算术指令 ·············· 67
- 9.3　小结 ·············· 69

第 10 章　堆栈 ·············· 71
- 10.1　理解堆栈 ·············· 71
- 10.2　跟踪堆栈 ·············· 74
- 10.3　小结 ·············· 76

第 11 章　浮点运算 ·············· 77
- 11.1　单精度与双精度 ·············· 77
- 11.2　浮点数编程 ·············· 78
- 11.3　小结 ·············· 81

第 12 章　函数 ·············· 83
- 12.1　编写一个简单的函数 ·············· 83
- 12.2　更多函数 ·············· 85
- 12.3　小结 ·············· 87

第 13 章　栈对齐和栈帧 ·············· 89
- 13.1　栈对齐 ·············· 89
- 13.2　有关栈帧的更多信息 ·············· 91
- 13.3　小结 ·············· 92

第 14 章　外部函数 ·············· 93
- 14.1　编译并链接函数 ·············· 93
- 14.2　扩展 makefile ·············· 97
- 14.3　小结 ·············· 98

第 15 章　调用约定 ·············· 99
- 15.1　函数参数 ·············· 100
- 15.2　栈布局 ·············· 103
- 15.3　保留寄存器 ·············· 106
- 15.4　小结 ·············· 107

第 16 章　位运算 ·············· 109
- 16.1　基础 ·············· 109
- 16.2　算术 ·············· 115
- 16.3　小结 ·············· 119

第 17 章　位操作 ·············· 121
- 17.1　修改位的其他方法 ·············· 121
- 17.2　位标志变量 ·············· 124
- 17.3　小结 ·············· 125

第 18 章　宏 ·············· 127
- 18.1　编写宏 ·············· 127
- 18.2　使用 objdump ·············· 129
- 18.3　小结 ·············· 130

第 19 章　控制台 I/O ·············· 131
- 19.1　使用 I/O ·············· 131
- 19.2　处理溢出 ·············· 133
- 19.3　小结 ·············· 137

第 20 章　文件 I/O ·············· 139
- 20.1　使用 syscall ·············· 139
- 20.2　文件处理 ·············· 140
- 20.3　条件汇编 ·············· 149
- 20.4　文件操作指令 ·············· 149
- 20.5　小结 ·············· 151

第 21 章　命令行 ·············· 153
- 21.1　访问命令行参数 ·············· 153
- 21.2　调试命令行 ·············· 154
- 21.3　小结 ·············· 156

第 22 章　从 C 到汇编 ·············· 157
- 22.1　编写 C 源文件 ·············· 157
- 22.2　编写汇编代码 ·············· 159
- 22.3　小结 ·············· 164

第 23 章　内联汇编 ·············· 165
- 23.1　基本内联汇编 ·············· 165
- 23.2　扩展内联汇编 ·············· 167
- 23.3　小结 ·············· 170

第 24 章　字符串 … 171
- 24.1　移动字符串 … 171
- 24.2　比较和扫描字符串 … 176
- 24.3　小结 … 181

第 25 章　cpuid … 183
- 25.1　使用 cpuid … 183
- 25.2　使用 test 指令 … 186
- 25.3　小结 … 188

第 26 章　SIMD … 189
- 26.1　标量数据和打包数据 … 189
- 26.2　数据对齐与不对齐 … 191
- 26.3　小结 … 192

第 27 章　小心 mxcsr … 193
- 27.1　操作 mxcsr 的位 … 194
- 27.2　分析程序 … 201
- 27.3　小结 … 202

第 28 章　SSE 对齐 … 203
- 28.1　未对齐示例 … 203
- 28.2　对齐示例 … 206
- 28.3　小结 … 210

第 29 章　SSE 打包整数 … 211
- 29.1　适用于整数的 SSE 指令 … 211
- 29.2　分析代码 … 213
- 29.3　小结 … 214

第 30 章　SSE 字符串操作 … 215
- 30.1　imm8 控制字节 … 216
- 30.2　使用 imm8 控制字节 … 217
  - 30.2.1　位 0 和 1 … 217
  - 30.2.2　位 2 和 3 … 217
  - 30.2.3　位 4 和 5 … 218
  - 30.2.4　位 6 … 218
  - 30.2.5　位 7 … 219
  - 30.2.6　标志 … 219
- 30.3　小结 … 220

第 31 章　搜索字符 … 221
- 31.1　确定字符串的长度 … 221
- 31.2　在字符串中搜索 … 224
- 31.3　小结 … 228

第 32 章　比较字符串 … 229
- 32.1　隐式长度 … 229
- 32.2　显式长度 … 232
- 32.3　小结 … 236

第 33 章　重排 … 237
- 33.1　重排初探 … 237
- 33.2　重排广播 … 243
- 33.3　重排反转 … 244
- 33.4　重排旋转 … 245
- 33.5　重排字节 … 245
- 33.6　小结 … 246

第 34 章　SSE 字符串掩码 … 247
- 34.1　搜索字符 … 247
- 34.2　搜索某个范围内的字符 … 253
- 34.3　搜索子字符串 … 258
- 34.4　小结 … 262

第 35 章　AVX … 263
- 35.1　测试是否支持 AVX … 263
- 35.2　AVX 程序示例 … 265
- 35.3　小结 … 270

第 36 章　AVX 矩阵运算 … 271
- 36.1　矩阵代码示例 … 271
- 36.2　矩阵打印：printm4x4 … 281
- 36.3　矩阵乘法：multi4x4 … 281
- 36.4　矩阵求逆：Inverse4x4 … 284
  - 36.4.1　Cayley-Hamilton 定理 … 284
  - 36.4.2　Leverrier 算法 … 285
  - 36.4.3　代码 … 286
- 36.5　小结 … 289

第 37 章　矩阵转置 291
37.1　转置代码示例 291
37.2　解包版本 295
37.3　重排版本 299
37.4　小结 301

第 38 章　性能调优 303
38.1　转置计算性能 303
38.2　迹计算性能 310
38.3　小结 317

第 39 章　你好，Windows 的世界 319
39.1　入门 319
39.2　编写一些代码 321
39.3　调试 323
39.4　syscall 323
39.5　小结 323

第 40 章　使用 Windows API 325
40.1　控制台输出 325
40.2　编译 Windows 程序 328
40.3　小结 330

第 41 章　Windows 中的函数 331
41.1　使用四个以上的参数 331
41.2　使用浮点数 337
41.3　小结 339

第 42 章　可变参数函数 341
42.1　Windows 中的可变参数函数 341
42.2　混合值 343
42.3　小结 345

第 43 章　Windows 文件 347
43.1　小结 350

后记 351

第 1 章

你的第一个程序

几代程序员都通过学习如何在电脑屏幕上显示"hello，world"来开始他们的编程生涯。这是一个传统，是从 Brian W. Kernighan 与 Dennis Ritchie 合著的一本书(*The C Programming Language*)开始的。Kernighan 在贝尔实验室开发了 C 编程语言。从那时起，C 语言发生了很大变化，但仍然是每个程序员都应该熟悉的语言。大多数"现代"和"流行"的编程语言都源于 C，C 有时被称为可移植的汇编语言。作为有抱负的汇编程序员，你应该熟悉 C。为了尊重这一传统，我们将从一个在屏幕上显示"hello, world"的汇编程序开始。代码清单 1-1 展示了"hello, world"程序的汇编语言版本，本章将对其进行分析。

代码清单 1-1：hello.asm

```
;hello.asm
section .data
    msg db "hello, world",0
section .bss
section .text
    global main
main:
    mov    rax, 1        ; 1 表示写入
    mov    rdi, 1        ; 1 表示标准输出
    mov    rsi, msg      ; 需要显示的是字符串存放在 rsi 中
    mov    rdx, 12       ; 字符串的长度，不包括 0
    syscall              ; 显示字符串
    mov    rax, 60       ; 60 表示退出
    mov    rdi, 0        ; 0 是成功退出代码
    syscall              ; 退出
```

1.1 编辑、汇编、链接和运行(或调试)

有许多优秀的文本编辑器，包括免费和商业的。寻找一款支持 64 位 NASM 语法突出显示功能的编辑器。大多数情况下，你需要下载某种插件或软件包才能突出显示语法。

注意 在本书中，我们将为 Netwide 汇编器(NASM)编写代码。还有其他汇编器，例如 Microsoft 的 YASM、FASM、GAS 或 MASM。与计算机世界中的一切技术一样，有时会激烈地讨论哪种汇编器是最好的。在本书中，我们将使用 NASM，因为它可以在 Linux、Windows 和 macOS 上使用，并且有一个很大的社区支持 NASM。你可以在 www.nasm.us 上找到相关手册。

我们将使用已安装了汇编语言语法文件的 gedit。gedit 是 Linux 的标准编辑器，我们将使用 Ubuntu Desktop 18.04.2 LTS。可以在 https://wiki.gnome.org/action/show/Projects/GtkSourceView/LanguageDefinitions 中找到语法突出显示文件。下载文件 asm-intel.lang，将其复制到 /usr/share/gtksourceview*.0/language specs/，然后将星号(*)替换为系统上安装的版本。当打开 gedit 时，可以在 gedit 窗口的底部选择编程语言，这里是 Assembler(Intel)。

在 gedit 屏幕上，代码清单 1-1 展示的 hello.asm 文件如图 1-1 所示。

```
1 ; hello.asm
2 section .data
3     msg db      "hello, world",0
4 section .bss
5 section .text
6     global main
7 main:
8     mov     rax, 1
9     mov     rdi, 1
10    mov     rsi, msg
11    mov     rdx, 12
12    syscall
13    mov     rax, 60
14    mov     rdi, 0
15    syscall
```

图 1-1　gedit 中的 hello.asm

语法突出显示使汇编代码更易于阅读。

编写汇编程序时，我们需要在屏幕上打开两个窗口：一个包含 gedit 的窗口，其中包含汇编程序源代码；另一个是位于项目目录中带有命令行提示符的窗口，以便我们可以轻松地在编辑和操作项目文件之间进行切换(汇编和运行程序、调试等)。对于更复杂和更大的项目，这是不可行的；你将需要一个集成的开发环境(IDE)。但是目前，使用简单的文本编辑器和命令行就可以了。这样做的好处是我们可以专注于汇编程序，而不是 IDE 的繁杂事项。在后续章节中，我们将讨论有用的工具和实用程序，其中一些具有图形用户界面，而另一些则面向命令行。但是，解释和使用 IDE 超出了本书的范围。

对于本书中的每个练习，我们都使用一个单独的项目目录，其中包含项目需要的和生成的所有文件。

当然，除了文本编辑器，还必须检查是否安装了许多其他工具，例如 GCC、GDB、make 和 NASM。首先需要 GCC，它是默认的 Linux 编译器链接器。

GCC 代表 GNU 编译器集合，是 Linux 上的标准编译器和链接器工具。GNU 表示 **GNU is Not Unix**，是递归的首字母缩写词。用递归的首字母缩写来命名事物是 LISP 程序员在 70 年代开始采取的做法，这是一个蹩脚的老笑话。

在命令行输入 gcc -v。如果安装了 GCC，它将以许多消息作为响应。如果尚未安装，请在命令行输入以下命令进行安装：

```
sudo apt install gcc
```

对 gdb-v 和 make-v 也执行同样的操作。如果你不理解这些命令，请在继续之前复习一下 Linux 相关知识。

你需要安装 NASM 和 build-essential，其中包含许多将要使用的工具。为此，请在 Ubuntu Desktop 18.04 中使用以下命令：

```
sudo apt install build-essential nasm
```

在命令行输入 nasm -v，如果正确安装了 nasm，它将以版本号响应。如果已经安装好了这些程序，就可以开始第一个汇编程序了。

将代码清单 1-1 所示的"hello, world"程序输入到你喜欢的编辑器中，并将其保存为 hello.asm。如前所述，使用单独的目录保存第一个项目的文件。我们将在本章后面解释每一行代码。请注意汇编源代码的以下特征("源代码"是保存刚刚输入的程序指令的 hello.asm 文件)：

- 在代码中，可以使用制表符(tab)、空格和换行符使代码更具可读性。
- 每行一条指令。
- 分号后的文字是注释，换句话说，是对人类有益的解释。计算机会忽略注释。

使用文本编辑器创建另一个包含代码清单 1-2 中各行的文件。

代码清单 1-2：hello.asm 的 makefile

```
#hello.asm 的 makefile
hello: hello.o
    gcc -o hello hello.o -no-pie
```

```
hello.o: hello.asm
    nasm -f elf64 -g -F dwarf hello.asm -l hello.lst
```

图 1-2 展示了 gedit 中的 makefile。

```
1 #makefile for hello.asm
2 hello: hello.o
3         gcc -o hello hello.o -no-pie
4 hello.o: hello.asm
5         nasm -f elf64 -g -F dwarf hello.asm -l hello.lst
```

图 1-2　gedit 中的 makefile

将此文件另存为 makefile(与 hello.asm 在同一目录中)，然后退出编辑器。

make 将使用 makefile 自动编译程序。编译程序意味着检查源代码中是否有错误，从操作系统添加所有必需的服务以及将代码转换为一系列机器可读指令。在本书中，我们将使用简单的 makefile。如果想了解有关 makefile 的更多信息，请参见以下手册：

https://www.gnu.org/software/make/manual/make.html

参考指南请见：

https://www.tutorialspoint.com/makefile/

你可以自下而上阅读 makefile，以查看其作用。这是一个简化的解释：make 实用程序与依赖关系树一起工作。它明确了 hello 依赖于 hello.o，hello.o 依赖于 hello.asm，而 hello.asm 则不依赖其他任何东西。make 比较 hello.asm 和 hello.o 的最后修改时间，如果 hello.asm 的时间较新，make 将在 hello.o 之后执行该行，即 hello.asm。然后 make 重新开始读取 makefile 并发现 hello.o 的修改时间比 hello 的时间晚。因此，它在 hello 之后执行该行，即 hello.o。

在 makefile 的最后一行，NASM 被用作汇编器。-f 后面是输出格式，在我们的例子中是 elf64，它表示 64 位的可执行和可链接格式。-g 表示我们希望以-F 选项之后指定的调试格式包含调试信息。我们使用 DWARF 调试格式。发明这种格式的软件极客似乎喜欢 Tolkien 撰写的《霍比特人》和《指环王》，所以也许这就是为什么他们认为 DWARF 是 ELF 的一个很好的补充。实际上 DWARF 是 Debug With Arbitrary Record Format 的缩写。

STABS 是另一种调试格式，与 Tolkien 小说中的所有刺伤(stab)无关。名称来自符号表字符串(Symbol Table Strings)。我们不会在本书使用 STABS。

-l 告诉 NASM 生成.lst 文件。我们将使用.lst 文件检查汇编的结果。NASM 将创建一个扩展名为.o 的目标文件。接下来，链接器将使用该目标文件。

注意 通常，NASM 会生成许多隐秘消息，并拒绝提供目标文件。有时，NASM 频繁报错会使你发疯。这种情况下，保持冷静，喝杯咖啡并检查代码非常重要，因为你可能做错了什么。随着汇编经验的增加，你会更快地发现错误。

当你最终"说服"NASM 为你生成一个目标文件时，该目标文件将通过链接器进行链接。链接器获取目标代码，并在系统中搜索所需的其他文件，通常是系统服务或其他目标文件。链接器将这些文件与你生成的目标代码组合在一起，并生成一个可执行文件。当然，链接器将在所有可能的情况下向你报告丢失的内容等。如果是这样，请再喝一杯咖啡，然后检查你的源代码和 makefile。

在我们的例子中，使用 GCC 的链接功能(此处重复以供参考)：

```
hello: hello.o
    gcc -o hello hello.o -no-pie
```

默认情况下，最新的 GCC 链接器和编译器会生成与位置无关的可执行文件(PIE，Position-Independent Executable)。这是为了防止黑客调查程序使用内存并最终干扰程序执行。在本例中，我们将不编译与位置无关的可执行文件。这确实会使我们程序的分析复杂化(出于安全原因，故意这样做)。因此，我们在 makefile 中添加参数-no-pie。

最后，可以通过使用井号(#)，在 makefile 中插入注释。

```
#makefile for hello.asm
```

我们之所以使用 GCC，是因为可以轻松地从汇编代码中访问 C 标准库函数。为了使生活更轻松，我们将时不时地使用 C 语言函数简化示例汇编代码。众所周知，Linux 上另一个流行的链接器是 ld，即 GNU 链接器。

如果前面的内容对你没有意义，请不要担心，喝杯咖啡然后继续吧；那只是背景信息，在此阶段并不重要。只要记住，makefile 是你的朋友，并为你做很多工作；此时你只需要确认无错误。

在命令提示符下，转到保存 hello.asm 文件和 makefile 的目录。输入 make 来汇编和生成程序，然后在命令提示符下输入./hello 来运行程序。如果你在命令提示符前看到消息 hello,world，则说明一切正常。否则，可能是出现了一些输入错误或其他错误，需要检查源代码或 makefile。再斟上一杯咖啡，祝你调试愉快！

图 1-3 展示了输出示例。

```
jo@UbuntuDesktop:~/Desktop/linux64/gcc/01 hello $
jo@UbuntuDesktop:~/Desktop/linux64/gcc/01 hello $
jo@UbuntuDesktop:~/Desktop/linux64/gcc/01 hello $ make
nasm -f elf64 -g -F dwarf hello.asm -l hello.lst
gcc -o hello hello.o
jo@UbuntuDesktop:~/Desktop/linux64/gcc/01 hello $ ./hello
hello, worldjo@UbuntuDesktop:~/Desktop/linux64/gcc/01 hello $
```

图 1-3　hello, world 的输出

1.2　汇编程序的结构

第一个程序说明了汇编程序的基本结构。以下是汇编程序的主要组成部分：

- .data 段
- .bss 段
- .text 段

1.2.1　.data 段

在 .data 段(section)中，使用以下格式声明和定义初始化数据：

<变量名称>　　<类型>　　<值>

如果 section .data 中包含变量，当源代码被汇编并链接到可执行文件时，将为该变量分配内存。变量名称是符号名称，对内存位置和变量的引用可以占用一个或多个内存位置。变量名称是指变量在内存中的起始地址。

变量名以字母开头，后跟字母、数字或特殊字符。表 1-1 列出了可能的数据类型。

表 1-1　数据类型

类型	长度	名称
db	8 位	Byte(字节)
dw	16 位	Word(字)
dd	32 位	Double word(双字)
dq	64 位	Quadword(四字)

在示例程序中，.data 段包含一个变量 msg，这是一个符号名称，指向内存地址 h，是字符串"hello, world"的第一个字节。因此，msg 指向字母 h，msg + 1 指向字母 e，以此类推。这个变量称为字符串，是字符的连续列表。字符串是内存中字符的"列表"或"数组"。实际上，内存中的任何连续列表都可以视为字符串；

字符可以是人类可读的，对于人类是有意义的。

使用 0 表示人类可读字符串的结尾很方便。你可以忽略使用终止 0，但后果自负。我们所指的终止 0 不是 ASCII 0，而是 0。它是一个数字零，并且 0 处的存储位置包含 8 个 0 位(bit)。如果你对 ASCII 的首字母缩写不理解，请自行使用搜索引擎进行了解。掌握 ASCII 的含义对于编程很重要。这里是简短的解释：供人类使用的字符对应于计算机中的特殊代码。大写字母 A 的代码为 65，B 的代码为 66，以此类推。换行符或新行的代码为 10，而 NULL 的代码为 0。因此，我们以 NULL 终止字符串。在命令行输入 man ascii 时，Linux 将显示一个 ASCII 表。

.data 段还可以包含常量，常量是程序中无法更改的值。它们的声明格式如下：

<常量名称>   equ   <值>

例如：

pi   equ   3.1416

1.2.2 .bss 段

首字母缩写词 bss 表示以符号开头的块(Block Started by Symbol)，它的历史可以追溯到 20 世纪 50 年代，当时是为 IBM 704 开发的汇编语言的一部分。这段用来存放未初始化的变量。未初始化变量的空间在此段中声明，格式如下：

<变量名称>   <类型>   <数字>

表 1-2 显示了可能的 bss 数据类型。

表 1-2 bss 数据类型

类型	长度	名称
resb	8 位	Byte(字节)
resw	16 位	Word(字)
resd	32 位	Double word(双字)
resq	64 位	Quadword(四字)

例如，以下内容声明了一个包含 20 个双字的数组：

dArray resd 20

.bss 段中的变量不包含任何值；这些值将在稍后的执行中分配。内存位置不是在编译时保留的，而是在执行时保留的。在以后的示例中，我们将展示.bss 段的用法。当程序开始执行时，程序会从操作系统中请求所需的内存，分配给 .bss 段中的变量并初始化为零。如果执行时没有足够的内存可用于 .bss 变量，程序将崩溃。

1.2.3　.txt 段

.txt 段是所有操作所在的位置。本节包含程序代码，并从以下内容开始：

```
        global main
main:
```

main: 部分称为标签。对于仅含有标签的行，最好在标签后跟一个冒号。否则，汇编器会发送警告。你不应该忽略这个警告！如果标签后面有其他指令，则不需要冒号，但是最好养成使所有标签都以冒号结尾的习惯。这样做可以提高代码的可读性。

在 hello.asm 代码中，在 main: 标签之后，准备了诸如 rdi、rsi 和 rax 的寄存器，以便在屏幕上输出消息。我们将在第 2 章中介绍有关寄存器的更多信息。在这里，我们将使用系统调用在屏幕上显示一个字符串。也就是说，我们将要求操作系统自动完成这项工作。

- 系统调用代码 1 被放入寄存器 rax 中，表示"写入"。
- 为将一些值放入寄存器，我们使用指令 mov。实际上，该指令不会移动任何东西；它将从源复制到目标。格式如下：

  ```
  mov destination, source
  ```

  - mov 指令的用法如下：
    - ◆ mov 寄存器, 即时值
    - ◆ mov 寄存器, 内存
    - ◆ mov 内存, 寄存器
    - ◆ 不合法：mov 内存, 内存
- 在代码中，用于写入的输出目标存储在寄存器 rdi 中，1 表示标准输出(在本例中，是输出到屏幕上)。
- 要显示的字符串的地址被放入寄存器 rsi。
- 我们在寄存器 rdx 中放置消息的长度。数一数 "hello, world" 包含多少个字符。不要计算字符串本身的引号或结尾的 0。如果算上结尾 0，程序将尝试显示一个 NULL 字节，这没有意义。

- 然后将执行系统调用(syscall)，字符串 msg 将显示在标准输出上。syscall 是对操作系统的函数调用。
- 为避免在程序完成时出现错误消息，需要清除程序出口。首先将 60 写入 rax，表示"退出"。接下来把"成功"退出代码 0 放入 rdi，然后执行系统调用。这样程序就会正常退出而不会报错。

系统调用会要求操作系统执行特定的操作。每个操作系统都有不同的系统调用参数列表，并且 Linux 的系统调用与 Windows 或 macOS 不同。在本书中，我们使用适合 64 位汇编的 Linux 系统调用。你可在 https://blog.rchapman.org/posts/Linux_System_Call_Table_for_x86_64/ 找到更多信息。

请注意，32 位系统调用与 64 位系统调用不同。读取代码时，请始终验证代码是针对 32 位还是 64 位系统编写的。

转到命令行模式，然后查找文件 hello.lst。如 makefile 中所指定，该文件是汇编期间，在链接之前生成的。在编辑器中打开 hello.lst，将看到汇编代码清单；在最左侧的列中，将看到代码的相对地址，在下一列中，将看到代码翻译成机器语言(十六进制)。图 1-4 展示了 hello.lst。

```
1                              section .data
2   00000000 68656C6C6F2C20776F-    msg db      "hello, world",0
3   00000009 726C6400
4                              section .bss
5                              section .text
6                                   global main
7                              main:
8   00000000 B801000000             mov     rax, 1      ; 1 = write
9   00000005 BF01000000             mov     rdi, 1      ; 1 = to stdout
10  0000000A 48BE-                  mov     rsi, msg    ; string to display in rsi
11  0000000C [0000000000000000]
12  00000014 BA0C000000             mov     rdx, 12     ; length of the string, without 0
13  00000019 0F05                   syscall             ; display the string
14  0000001B B83C000000             mov     rax, 60     ; 60 = exit
15  00000020 BF00000000             mov     rdi, 0      ; 0 = success exit code
16  00000025 0F05                   syscall             ; quit
```

图 1-4 hello.lst

第一列是行号，第二列是八位数字。这列表示内存位置。当汇编程序编译对象文件时，它还不知道将使用哪个内存位置。所以，它从不同部分的位置 0 开始。.bss 段不分配内存。

我们在第二列中看到将汇编指令转换为十六进制代码的结果。例如，将 mov rax 转换为 B8，将 mov rdi 转换为 BF。这些是机器指令的十六进制表示。另外请注意，msg 字符串转换为十六进制 ASCII 字符。稍后，你将了解有关十六进制表示法的更多信息。要执行的第一条指令从地址 00000000 开始，并占用 5 个字节：B8 01 00 00 00。有两个双零(00)用于填充和内存对齐。内存对齐是汇编器和编译器用来优化代码的功能。可为汇编程序和编译器提供不同的标志，以获得尽可能小的可执行文件、最快的代码或二者的组合。在后续章节中，我们将讨论优化，以提高执行速度。

下一条指令从地址 00000005 开始，以此类推。内存地址有八位数字(即 8 个字节)；每个字节有 8 位。因此，地址有 64 位；确实，我们使用的是 64 位汇编器。看看 msg 是如何被引用的。由于尚不知道 msg 的存储位置，因此它被标记为 [0000000000000000]。

显而易见，汇编器助记符和符号名称比十六进制值更容易记住，要知道有数百种助记符，并且有多个操作数，每个操作数会产生更多的十六进制指令。在计算机的早期，程序员使用机器语言，即第一代编程语言。汇编语言是第二代编程语言，具有"易于记忆"的助记符。

1.3 小结

本章内容：
- 汇编程序的基本结构，包括不同的组成部分
- 内存，带有符号名称的地址
- 寄存器
- 汇编指令：mov
- 如何使用系统调用
- 机器代码和汇编代码之间的区别

第 2 章

二进制数、十六进制数和寄存器

在当今的计算机世界中，位(bit)是最小的信息片段。一个位的值可以是 1 或 0。在本章中，我们将研究如何组合位以表示数据，例如整数或浮点值。对人类来说直观的十进制表示形式对于计算机而言并不是理想选择。当你有一个只有两个可能值(1 或 0)的二进制系统时，使用 2 的幂进行运算将更有效。当我们谈论历史上的计算机时代时，有 8 位 CPU(2^3)、16 位 CPU(2^4)、32 位 CPU(2^5)，目前大多数是 64 位 CPU(2^6)。但对人类来说，处理 1 和 0 的长字符串是不切实际的，甚至是不可能的。在本章中，我们将展示如何将位转换为更易使用的十进制或十六进制值。之后，我们将讨论寄存器和数据存储区域，这些区域可帮助处理器执行逻辑和算术指令。

2.1 二进制简短课程

计算机使用二进制数字(0 和 1)完成工作。分组在一起的八个二进制数字称为一个字节。但是，二进制数字对于人类来说太长了，更不用说要记住了。十六进制数更易于使用(只是稍微方便一点)，因为每个 8 位字节只能由两个十六进制数表示。

如果要以其他显示格式查看二进制、十进制或十六进制值，则需要使用转换器。互联网上有很多转换计算器。以下是一些易于使用的方法：

- www.binaryconvert.com
- https://www.binaryhexconverter.com
- https://babbage.cs.qc.cuny.edu/IEEE-754/

表 2-1 是基本的转换表，记住此表将很有帮助。

表 2-1 基本的转换表

十进制	十六进制	二进制
0	0	0000
1	1	0001
2	2	0010
3	3	0011
4	4	0100
5	5	0101
6	6	0110
7	7	0111
8	8	1000
9	9	1001
10	a	1010
11	b	1011
12	c	1100
13	d	1101
14	e	1110
15	f	1111

2.1.1 整数

有两种整数，有符号和无符号。有符号整数的最左位设置为 1(如果为负)或 0(如果为正)。无符号整数为 0 或正数；没有符号位的空间。为了能够进行整数运算，负整数使用二进制补码表示。负数的二进制表示如下所示：

(1) 写出绝对值的二进制。
(2) 取补码(将所有 1 更改为 0，将 0 更改为 1)。
(3) 加 1。

这是一个使用 16 位数字而不是 64 位数字的示例(以使示例易于管理)：

```
十进制数字    =    17
二进制数字    =    0000 0000 0001 0001
十六进制数字  =    0    0    1    1    =11
十进制数字    =    -17
二进制数的绝对值 = 0000 0000 0001 0001
补码 =              1111 1111 1110 1110
加 1=               1111 1111 1110 1111
```

第 2 章 ■ 二进制数、十六进制数和寄存器

```
十六进制 =              f    f    e    f    = ffef
验证：   -17  11111111       11101111
加 (+)：  +17  00000000       00010001
等于：    0   00000000       00000000
```

为了与十进制数区分，十六进制数前面通常以 0x 开头，因此十六进制中的-17 是 0xffef。如果你研究一个机器语言列表，一个.lst 文件，并且看到数字 0xffef，那么你必须从上下文中找出它是有符号整数还是无符号整数。如果它是一个有符号整数，则表示十进制的-17。如果它是无符号整数，则表示 65519。当然，如果它是内存地址，则它是无符号的(你明白吗？)。有时，你还会在汇编代码中看到其他符号，例如，0800h 是十六进制数，10010111b 是二进制数，420o 是八进制数。是的，也可以使用八进制数。在为文件 I/O 编写代码时，我们将使用八进制数。如果你需要转换整数，使用前面提到的网站即可。

2.1.2 浮点数

根据 IEEE-754 标准，浮点数使用二进制或十六进制形式书写。这个过程甚至比使用整数还复杂；如果你想知道细节，可以从这里开始：

http://mathcenter.oxford.emory.edu/site/cs170/ieee754/

同样，如果需要转换浮点数，请使用前面提到的网站；我们将不在这里做进一步介绍。

2.2 寄存器简短课程

CPU 是计算机的大脑，它通过全面使用计算机的寄存器和内存来执行程序指令，并对这些寄存器和内存进行数学和逻辑运算。因此，具有寄存器和内存的基本知识并了解如何使用它们很重要。这里简要介绍寄存器。有关寄存器使用的更多信息将在后续章节中阐明。寄存器是 CPU 用来存储数据、指令或内存地址的存储位置。寄存器数量很少，但是 CPU 可以非常快速地读取和写入它们。你可将寄存器视为一种暂存器，供处理器存储临时信息。如果速度很重要，要记住的一条规则是 CPU 访问寄存器的速度比访问内存的速度快得多。

如果你对此一无所知，也不必担心。当我们在接下来的章节中使用寄存器时，一切都会变得明了。

2.2.1 通用寄存器

有 16 个通用寄存器，每个寄存器可以用作 64 位、32 位、16 位或 8 位寄存器。在表 2-2 中，你可以看到不同大小的每个寄存器的名称。四个寄存器 rax、rbx、rcx 和 rdx 可以有两种 8 位寄存器：低 8 位(16 位寄存器的下半部分)和高 8 位(16 位寄存器的上半部分)。

表 2-2　寄存器名称

64 位	32 位	16 位	低 8 位	高 8 位	备注
rax	eax	ax	al	ah	
rbx	ebx	bx	bl	bh	
rcx	ecx	cx	cl	ch	
rdx	edx	dx	dl	dh	
rsi	esi	si	sil	-	
rdi	edi	di	dil	-	
rbp	ebp	bp	bpl	-	基指针
rsp	esp	sp	spl	-	栈指针
r8	r8d	r8w	r8b	-	
r9	r9d	r9w	r9b	-	
r10	r10d	r10w	r10b	-	
r11	r11d	r11w	r11b	-	
r12	r12d	r12w	r12b	-	
r13	r13d	r13w	r13b	-	
r14	r14d	r14w	r14b	-	
r15	r15d	r15w	r15b	-	

尽管 rbp 和 rsp 被称为通用寄存器，但是应谨慎处理它们，因为它们在程序执行期间由处理器使用。我们将在更高级的章节中大量使用 rbp 和 rsp。

64 位寄存器包含一组 64 位 0 和/或 1，即 8 字节。当我们在 hello,world 程序中将 60 放入 rax 时，rax 包含以下内容：

00000000 00000000 00000000 00000000 00000000 00000000 00000000 00111100

这是 64 位寄存器中数字 60 的二进制表示。

32 位寄存器是 64 位寄存器的 32 个低(最右)位的集合。同样，一个 16 位寄存器和一个 8 位寄存器分别由 64 位寄存器的最低 16 位和最低 8 位组成。

> 请记住,"低"位始终是最右边的位。

位号 0 是最右边的位;我们从右边开始计数,并从索引 0(而不是 1)开始。因此,一个 64 位寄存器的最左位的索引是 63,而不是 64。

因此,当 rax 的值为 60 时,也可以说 eax 现在包含以下内容:

00000000 00000000 00000000 00111100

或 ax 包含以下内容:

00000000 00111100

或 al 包含以下内容:

00111100

2.2.2 指令指针寄存器(rip)

处理器通过将下一条指令的地址存储在 rip 中来跟踪要执行的下一条指令。你可以自行将 rip 的值更改为所需的值;但你不应该那样做。更改 rip 中值的一种更安全的方法是使用跳转指令。我们将在后续章节讨论。

2.2.3 标志寄存器(Flag Register)

表 2-3 是标志寄存器 rflags 的布局。执行指令后,程序可以检查是否设置了某个标志(例如,ZF=1),然后采取相应的动作。

表 2-3 标志寄存器

名称	标记	位	内容
Carry	CF	0	先前的指令有一个进位
Parity	PF	2	最后一个字节有偶数个 1
Adjust	AF	4	BCD 操作
Zero	ZF	6	上一条指令的结果为零
Sign	SF	8	上一条指令的最高有效位等于 1
Direction	DF	10	字符串操作的方向(递增或递减)
Overflow	OF	11	上一条指令导致溢出

我们将在本书中解释和使用许多标记。

还有一个称为 MXCSR 的标志寄存器,将在第 26 章使用;我们将在那里详细解释 MXCSR。

2.2.4　xmm 和 ymm 寄存器

这些寄存器用于浮点计算和 SIMD。稍后我们将从浮点指令开始广泛使用 xmm 和相应的 ymm 寄存器。

除了前面介绍的寄存器外，还有很多寄存器，但本书中将不再使用其他寄存器。现在暂时将理论搁置一边，是时候开始真正的工作了！

2.3　小结

本章内容：
- 如何以十进制、二进制和十六进制格式表示值
- 如何使用寄存器和标志

第 3 章

用调试器进行程序分析：GDB

本章将介绍如何调试汇编程序。调试是一项重要的技能，因为可以通过使用调试器研究十六进制、二进制或十进制表示形式的寄存器和内存的内容。在上一章你已经知道 CPU 会大量使用寄存器和内存，调试器使你可以在逐步执行指令的同时查看寄存器、内存和标志的内容如何变化。也许你已经体验到你的第一个汇编程序在执行时由于诸如"内存分段错误"的不友好消息而崩溃。使用调试器，你可以单步执行程序，并找出问题出在哪里以及为什么出错。

3.1 开始调试

汇编并链接 hello, world 程序后，如果没有错误，你将获得一个可执行文件。使用调试器工具，可将可执行程序加载到计算机内存中，并在检查各种寄存器和内存位置的同时逐行执行该程序。有几种免费的和商业的调试器可供选择。在 Linux 中，所有调试器的核心都是 GDB。它是一个命令行程序，具有非常晦涩难懂的命令。在后续章节中，我们将使用基于 GDB 的具有图形用户界面的工具 SASM。然而，了解 GDB 本身的基本知识可能会很有用，因为在 SASM 中，并非所有 GDB 功能都可用。

在以后的汇编程序员生涯中，你肯定会看到各种具有良好用户界面的调试器，每种调试器都针对特定的平台，如 Windows、Mac 或 Linux。与命令行调试器相比，这些图形界面调试器将帮助你轻松调试长而复杂的程序。但是 GDB 是进行 Linux 调试的全面而"快速且猥琐的"方式。大多数 Linux 开发系统上都默认安装了 GDB，如果没有安装，也可以轻松安装 GDB 并进行调试，而不会给系统带来太多开销。我们现在将使用 GDB 为你提供一些基本知识，并在后续章节中使用其他工具。值得注意的是，GDB 似乎是为调试高级语言开发的。调试汇编语言程序时，某些特性没有任何帮助。

第一次使用命令行调试器调试程序可能让人不知所措。阅读本章时不要绝望；你会发现，随着进一步学习，事情会变得越来越容易。

要开始调试 hello 程序，请在命令行中导航到保存 hello 程序的目录。在命令行提示符下，输入以下内容：

```
gdb hello
```

GDB 会将可执行的 hello 加载到内存中，并用自己的提示符(gdb)应答，等待你的指令。如果输入以下内容：

```
list
```

GDB 将显示多行代码。再次输入 list，GDB 将显示下一些行，以此类推。要列出特定的行(例如，代码的开头)，请输入 list 1。图 3-1 展示了一个示例。

```
jo@UbuntuDesktop:~/Desktop/linux64/gcc/01 hello $ gdb hello
GNU gdb (Ubuntu 7.11.1-0ubuntu1~16.5) 7.11.1
Copyright (C) 2016 Free Software Foundation, Inc.
License GPLv3+: GNU GPL version 3 or later <http://gnu.org/licenses/gpl.html>
This is free software: you are free to change and redistribute it.
There is NO WARRANTY, to the extent permitted by law.  Type "show copying"
and "show warranty" for details.
This GDB was configured as "x86_64-linux-gnu".
Type "show configuration" for configuration details.
For bug reporting instructions, please see:
<http://www.gnu.org/software/gdb/bugs/>.
Find the GDB manual and other documentation resources online at:
<http://www.gnu.org/software/gdb/documentation/>.
For help, type "help".
Type "apropos word" to search for commands related to "word"...
Reading symbols from hello...done.
(gdb) list
1           section .data
2               msg db      "hello, world",0
3           section .bss
4           section .text
5               global main
6           main:
7               mov     rax, 1      ; 1 = write
8               mov     rdi, 1      ; 1 = to stdout
9               mov     rsi, msg    ; string to display in rsi
10              mov     rdx, 12     ; length of the string, without 0
(gdb)
```

图 3-1　GDB list 输出

如果你的屏幕上的输出与此处显示的不同(包含许多％符号)，是因为你的 GDB 配置为使用 AT＆T 语法。我们将使用英特尔语法风格，将展示如何在一分钟内改变风格。

如果输入以下命令：

```
run
```

GDB 将运行你的 hello 程序，打印 hello,world，并返回其提示符(gdb)。显示的结果如图 3-2 所示。

```
jo@UbuntuDesktop:~/Desktop/linux64/gcc/01 hello $ gdb hello
GNU gdb (Ubuntu 7.11.1-0ubuntu1~16.5) 7.11.1
Copyright (C) 2016 Free Software Foundation, Inc.
License GPLv3+: GNU GPL version 3 or later <http://gnu.org/licenses/gpl.html>
This is free software: you are free to change and redistribute it.
There is NO WARRANTY, to the extent permitted by law.  Type "show copying"
and "show warranty" for details.
This GDB was configured as "x86_64-linux-gnu".
Type "show configuration" for configuration details.
For bug reporting instructions, please see:
<http://www.gnu.org/software/gdb/bugs/>.
Find the GDB manual and other documentation resources online at:
<http://www.gnu.org/software/gdb/documentation/>.
For help, type "help".
Type "apropos word" to search for commands related to "word"...
Reading symbols from hello...done.
(gdb) run
Starting program: /home/jo/Desktop/linux64/gcc/01 hello /hello
hello, world[Inferior 1 (process 4698) exited normally]
(gdb)
```

图 3-2　GDB run 输出

输入 quit 即可退出 GDB。

下面用 GDB 做一些有趣的事情！

但首先我们将更改反汇编风格；只有在上一个练习中有%符号时才执行此操作。如果可执行文件 hello 尚不存在，请将其加载到 GDB 中。

输入如下命令：

```
set disassembly-flavor intel
```

这会将反汇编的代码更改为一种已经熟悉的格式。你可以通过在 Linux shell 配置文件中使用适当的设置，使 Intel 成为 GDB 的默认风格。请参阅 Linux 发行版的文档。在 Ubuntu 18.04 中，在你的主目录中创建一个.gdbinit 文件，其中包含先前的 set 指令。注销并登录，从现在开始，你应该使用具有 Intel 风格的 GDB。

用 hello 启动 GDB 以开始分析。如你之前所学，hello,world 程序首先在 section.data 和 section.bss 中初始化一些数据，然后继续执行 main 标签。这是动作开始的地方，所以下面从这里开始检查。

在(gdb)提示符下输入如下命令：

```
disassemble main
```

GDB 或多或少地返回源代码。返回的源代码与你最初编写的源代码不完全相同。很奇怪，不是吗？这里发生了什么？需要执行一些分析。

图 3-3 展示了 GDB 在我们的计算机上返回的内容。

```
(gdb) disassemble main
Dump of assembler code for function main:
   0x00000000004004e0 <+0>:     mov    eax,0x1
   0x00000000004004e5 <+5>:     mov    edi,0x1
   0x00000000004004ea <+10>:    movabs rsi,0x601030
   0x00000000004004f4 <+20>:    mov    edx,0xc
   0x00000000004004f9 <+25>:    syscall
   0x00000000004004fb <+27>:    mov    eax,0x3c
   0x0000000000400500 <+32>:    mov    edi,0x0
   0x0000000000400505 <+37>:    syscall
   0x0000000000400507 <+39>:    nop    WORD PTR [rax+rax*1+0x0]
End of assembler dump.
(gdb)
```

图 3-3 GDB 反汇编输出

左侧的长串数字(从 0x00 ...开始)是内存地址，是程序的机器指令的存储位置。如你所见，从第二行的地址和<+5>开始，第一条指令 mov eax, 0x1 需要五个字节的内存。但是等一下，源代码是 mov rax,1，而 eax 是怎么回事呢？

如果查看第 2 章的寄存器表，就会发现 eax 是 rax 寄存器的低 32 位部分。汇编器很聪明，可以发现 64 位寄存器浪费了太多资源来存储数字 1，因此它使用 32 位寄存器。使用 edi 和 edx 代替 rdi 和 rdx 时也是如此。

64 位汇编器是 32 位汇编器的扩展，你将看到汇编器会尽可能使用 32 位指令。0x1 是十进制数字 1 的十六进制表示形式，0xd 是十进制数字 13，而 0x3c 是十进制数字 60。

msg 发生了什么？指令 mov rsi,msg 被 movabs rsi,0x601030 取代。现在不必考虑 movabs，它是由于 64 位寻址而存在的，用于将即时数(值)放入寄存器中。0x601030 是 msg 存储在计算机上的内存地址。根据你的情况，这可能是一个不同的地址。

在(gdb)提示符下，输入以下内容：

```
x/s 0x601030 (或 x/s 内存地址)
```

GDB 的响应如图 3-4 所示。

```
(gdb) x/s 0x601030
0x601030 <msg>: "hello, world"
(gdb)
```

图 3-4 GDB 输出 1

x 代表"检查"，s 代表"字符串"。GDB 响应 0x601030 是字符串 msg 的开始，并尝试显示整个字符串，直到字符串终止 0。现在你应该明白为什么我们在 hello,world 后面加一个终止 0 了。

也可以输入以下内容：

```
x/c 0x601030
```

会得到如图 3-5 所示的输出。

```
(gdb) x/c 0x601030
0x601030 <msg>:  104 'h'
(gdb)
```

图 3-5 GDB 输出 2

使用 c 表示你需要输出一个字符。GDB 在此返回 msg 的第一个字符，后跟该字符的十进制 ASCII 码。在 Google 上搜索一个 ASCII 码表以进行验证，并将其保存在便于将来使用的位置；不需要记住它。或打开其他终端窗口，然后在命令行输入 man ascii 以显示 ASCII 码表。

下面看一些其他例子。

使用如下命令显示从内存地址开始的 13 个字符(请参见图 3-6)：

```
x/13c 0x601030
```

```
(gdb) x/13c 0x601030
0x601030 <msg>:  104 'h'  101 'e'  108 'l'  108 'l'  111 'o'  44 ','  32 ' '  119 'w'
0x601038:        111 'o'  114 'r'  108 'l'  100 'd'  0 '\000'
(gdb)
```

图 3-6 GDB 输出 3

使用以下命令以十进制表示形式显示从内存地址开始的 13 个字符(请参见图 3-7)：

```
x/13d 0x601030
```

```
(gdb) x/13d 0x601030
0x601030 <msg>:  104      101      108      108      111      44       32       119
0x601038:        111      114      108      100      0
(gdb)
```

图 3-7 GDB 输出 4

使用以下命令，以十六进制表示形式显示从内存地址开始的 13 个字符(请参见图 3-8)：

```
x/13x 0x601030
```

```
(gdb) x/13x 0x601030
0x601030 <msg>:    0x68    0x65    0x6c    0x6c    0x6f    0x2c    0x20    0x77
0x601038:          0x6f    0x72    0x6c    0x64    0x00
(gdb)
```

图 3-8　GDB 输出 5

使用以下命令显示 msg(见图 3-9):

```
x/s &msg
```

```
(gdb) x/s &msg
0x601030 <msg>:  "hello, world"
(gdb)
```

图 3-9　GDB 输出 6

下面回到反汇编代码清单。输入以下内容:

```
x/2x 0x004004e0
```

这是以十六进制表示的从 0x004004e0 开始的两个内存地址的内容(见图 3-10)。

```
(gdb) x/2x 0x004004e0
0x4004e0 <main>:         0xb8    0x01
(gdb)
```

图 3-10　GDB 输出 7

这是我们用机器语言编写的第一条指令 mov eax,0x1。我们在检查 hello.lst 文件时看到了相同的指令。

3.2　继续进步

下面使用调试器逐步调试程序。别忘记首先需要在 GDB 中加载程序。

首先在程序中设置一个断点,暂停执行并允许我们检查一个数字或其他东西。输入以下内容:

```
break main
```

在我们的例子中,GDB 的响应输出如图 3-11 所示。

```
(gdb) break main
Breakpoint 1 at 0x4004e0: file hello.asm, line 7.
(gdb)
```

图 3-11　GDB 输出 8

然后输入以下内容:

第 3 章 ■ 用调试器进行程序分析：GDB

```
run
```

图 3-12 展示了输出结果。

```
(gdb) run
Starting program: /home/jo/Desktop/linux64/gcc/01 hello /hello

Breakpoint 1, main () at hello.asm:8
8           mov       rax, 1            ; 1 = write
(gdb)
```

图 3-12　GDB 输出 9

调试器在断点处停止，并显示将要执行的下一条指令。也就是说，尚未执行 mov rax,1。

输入以下内容：

```
info registers
```

GDB 返回的输出如图 3-13 所示。

```
(gdb) info registers
rax            0x4004e0    4195552
rbx            0x0         0
rcx            0x0         0
rdx            0x7fffffffddd8    140737488346584
rsi            0x7fffffffddc8    140737488346568
rdi            0x1         1
rbp            0x400510    0x400510 <__libc_csu_init>
rsp            0x7fffffffdce8    0x7fffffffdce8
r8             0x400580    4195712
r9             0x7ffff7de7ab0    140737351940784
r10            0x846       2118
r11            0x7ffff7a2d740    140737348032320
r12            0x4003e0    4195296
r13            0x7fffffffddc0    140737488346560
r14            0x0         0
r15            0x0         0
rip            0x4004e0    0x4004e0 <main>
eflags         0x246       [ PF ZF IF ]
cs             0x33        51
ss             0x2b        43
ds             0x0         0
es             0x0         0
fs             0x0         0
---Type <return> to continue, or q <return> to quit---
```

图 3-13　GDB 输出 10

除了指令指针 rip，寄存器的内容现在并不重要。寄存器 rip 的值为 0x4004e0，它是下一条要执行的指令的内存地址。检查反汇编代码清单：0x4004e0(在我们的示例中)指向第一条指令 mov rax,1。GDB 会在该指令之前停止并等待你的命令。一定**要记住 rip 所指向的指令尚未执行。**

在你的屏幕上，GDB 显示的内容可能与 0x4004e0 不同。没关系，它是内存中该特定行的地址，具体地址取决于你的计算机配置。

23

输入以下内容以前进一步：

step

输入以下内容，这是 info registers 的缩写：

i r

图 3-14 显示了输出内容。

```
(gdb) step
9           mov     rdi, 1              ; 1 = to stdout
(gdb) i r
rax            0x1      1
rbx            0x0      0
rcx            0x0      0
rdx            0x7fffffffddd8   140737488346584
rsi            0x7fffffffddc8   140737488346568
rdi            0x1      1
rbp            0x400510 0x400510 <__libc_csu_init>
rsp            0x7fffffffdce8   0x7fffffffdce8
r8             0x400580 4195712
r9             0x7ffff7de7ab0   140737351940784
r10            0x846    2118
r11            0x7ffff7a2d740   140737348032320
r12            0x4003e0 4195296
r13            0x7fffffffddc0   140737488346560
r14            0x0      0
r15            0x0      0
rip            0x4004e5 0x4004e5 <main+5>
eflags         0x246    [ PF ZF IF ]
cs             0x33     51
ss             0x2b     43
ds             0x0      0
es             0x0      0
fs             0x0      0
gs             0x0      0
(gdb)
```

图 3-14　GDB 输出 11

实际上，rax 现在包含 0x1，而 rip 包含下一条要执行的指令的地址。

进一步执行该程序，注意 rsi 如何接收 msg 的地址，在屏幕上打印 hello,world，然后退出。还请注意，rip 如何每次都指向下一条要执行的指令。

3.3　其他 GDB 命令

break 或 **b**：设置断点(正如之前所做的)。

disable breakpoint *number*
enable breakpoint *number*
delete breakpoint *number*

continue 或 **c**：继续执行，直到下一个断点。
step 或 **s**：进入当前行，最终跳转到被调用的函数。

next 或 n：跨过当前行，然后在下一行处停止。
help 或 h：显示帮助。
tui enable：启用简单的文本用户界面；请使用 tui disable 禁用。
print 或 p：打印变量、寄存器等的值。下面列举几个例子。
- 打印 **rax**　　p $rax
- 以二进制形式打印 **rax**　　p/t $rax
- 以十六进制形式打印 **rax**　　p/x $rax

要正确使用 GDB，必须在代码中插入函数序言(function prologue)和函数尾声(function epilogue)。我们将在下一章演示如何做到这一点；下一章讨论堆栈框架时，将讨论函数序言和函数尾声。诸如 hello,world 的简短程序是不需要考虑这个问题的。但是对于更复杂的程序，如果没有序言或尾声，GDB 将出现意外的行为。

尝试使用 GDB，参考联机手册(在命令行输入 man gdb)，并熟悉 GDB，因为即使使用 GUI 调试器，某些功能也可能不可用。或者，你可能根本不想在系统上安装 GUI 调试器。

3.4　稍加改进的 hello, world 程序

你应该已经注意到在打印 hello, world 之后，命令行提示符出现在同一行上。我们希望在单独的行上打印 hello, world，并在新行上显示命令行提示符。

代码清单 3-1 显示了实现代码。

代码清单 3-1：改进后的 hello, world 版本

```
;hello2.asm
section .data
    msg     db      "hello, world",0
    NL      db      0xa     ; 表示新行的 ascii 编码
section .bss
section .text
    global main
main:
    mov     rax, 1          ; 1 表示写入
    mov     rdi, 1          ; 1 表示标准输出
    mov     rsi, msg        ; 需要显示的字符串
    mov     rdx, 12         ; 字符串的长度,不包括 0
    syscall                 ; 显示字符串
    mov     rax, 1          ; 1 表示写入
    mov     rdi, 1          ; 1 表示标准输出
```

```
        mov     rsi, NL          ; 显示新行
        mov     rdx, 1           ; 字符串的长度
        syscall                  ; 显示字符串
        mov     rax, 60          ; 60 表示退出
        mov     rdi, 0           ; 0 表示成功退出代码
        syscall                  ; 退出
```

在编辑器中输入这些代码,并将其在新目录中另存为 hello2.asm。将先前的 makefile 复制到该新目录;在此 makefile 中,将 hello 的每个实例更改为 hello2 并保存文件。

我们添加了一个变量 NL,其中包含十六进制的 0xa,这是换行的 ASCII 码,并在打印 msg 之后立即打印此 NL 变量。继续汇编并运行它(见图 3-15)。

```
jo@UbuntuDesktop:~/Desktop/linux64/gcc/02 hello2$ make
nasm -f elf64 -g -F dwarf hello2.asm -l hello2.lst
gcc -o hello2 hello2.o
jo@UbuntuDesktop:~/Desktop/linux64/gcc/02 hello2$ ./hello2
hello, world
jo@UbuntuDesktop:~/Desktop/linux64/gcc/02 hello2$
```

图 3-15 改进后的 hello, world 版本

另一种方法是更改 msg,如下所示:

msg   db   "hello, world", 10, 0

10 是换行(新行)的十进制表示形式(十六进制为 0xa)。不要忘记将 rdx 增加到 13。

代码清单 3-2 展示了代码。将此文件在单独的目录中另存为 hello3.asm,复制并适当修改 makefile,然后编译并运行。

代码清单 3-2:另一种形式的代码

```
;hello3.asm
section .data
    msg db "hello, world",10,0
section .bss
section .text
    global main
main:
    mov     rax, 1           ; 1 表示写入
    mov     rdi, 1           ; 1 表示标准输出
    mov     rsi, msg         ; 需要显示的字符串
    mov     rdx, 13          ; 字符串的长度,不包括 0
    syscall                  ; 显示字符串
    mov     rax, 60          ; 60 表示退出
```

```
mov     rdi, 0          ; 0 表示成功退出代码
syscall                 ; 退出
```

在此版本中，换行是字符串的一部分；当然并非总是如此，因为换行是一种格式化指令，你可能只打算在显示字符串时使用，而不是在执行其他字符串处理函数时使用。但是，它可以使你的代码更简洁。这由你决定！

3.5 小结

本章内容：

- 如何使用命令行调试器(GDB)
- 如何打印一个新行(换行)

第 4 章

你的下一个程序：Alive and Kicking

现在，你已经对 GDB 有了充分了解，并且知道了汇编程序是什么样子，下面增加一些复杂性。在本章中，我们将展示如何获取字符串变量的长度。将展示如何使用 printf 打印整数和浮点数，帮助你进一步了解 GDB 命令。

代码清单 4-1 包含示例代码，我们将使用示例代码说明如何获得字符串的长度以及如何将数字值存储在内存中。

代码清单 4-1：alive.asm

```
;alive.asm
section .data
    msg1    db      "Hello, World!",10,0    ; 带有 NL 和终止 0 的字符串
    msg1Len equ $-msg1-1     ; 计算字符串的长度，需要减去 1(终止 0)
    msg2    db      "Alive and Kicking!",10,0 ; 带有 NL 和终止 0 的字符串
    msg2Len equ $-msg2-1     ; 计算字符串的长度，需要减去 1(终止 0)
    radius  dq      357      ; 没有字符串，不可显示？
    pi      dq      3.14     ; 没有字符串，不可显示？
section .bss
section .text
    global main
main:
    push    rbp              ; 函数序言
    mov     rbp,rsp          ; 函数序言
    mov     rax, 1           ; 1 表示写入
    mov     rdi, 1           ; 1 表示标准输出
    mov     rsi, msg1        ; 需要显示的字符串
    mov     rdx, msg1Len     ; 字符串的长度
    syscall                  ; 显示字符串
    mov     rax, 1           ; 1 表示写入
```

```
        mov         rdi, 1              ; 1 表示标准输出
        mov         rsi, msg2           ; 需要显示的字符串
        mov         rdx, msg2Len        ; 字符串的长度
        syscall                         ; 显示字符串
        mov         rsp,rbp             ; 函数尾声
        pop         rbp                 ; 函数尾声
        mov         rax, 60             ; 60 表示退出
        mov         rdi, 0              ; 0 表示成功退出代码
        syscall                         ; 退出
```

将此程序输入编辑器并另存为 alive.asm。创建包含代码清单 4-2 中内容的 makefile。

代码清单 4-2：alive.asm 的 makefile

```
#alive.asm 的 makefile
alive: alive.o
    gcc -o alive alive.o -no-pie
alive.o: alive.asm
    nasm -f elf64 -g -F dwarf alive.asm -l alive.lst
```

保存文件并退出编辑器。

在命令行提示符输入 make 来汇编并生成程序，然后在命令行提示符输入./alive 来运行程序。如果看到提示符处显示图 4-1 所示的输出，则说明一切按计划进行；否则，表明存在一些输入错误或其他错误。调试愉快！

```
jo@UbuntuDesktop:~/Desktop/linux64/gcc/04 alive$ make
nasm -f elf64 -g -F dwarf alive.asm -l alive.lst
gcc -o alive alive.o -ggdb -no-pie
jo@UbuntuDesktop:~/Desktop/linux64/gcc/04 alive$ ./alive
Hello, World!
Alive and Kicking!
jo@UbuntuDesktop:~/Desktop/linux64/gcc/04 alive$
```

图 4-1　alive.asm 的输出

4.1　alive 程序分析

在第一个程序 hello.asm 中，在 rdx 中放入 msg 的长度(13 个字符)，以便显示 msg。在 alive.asm 中，我们用了一个不错的功能计算变量长度，如下所示：

```
msg1Len equ $-msg1-1
```

$-msg1-1 部分的含义是获取该内存位置($)并减去 msg1 的内存位置。结果是 msg1 的长度。该长度-1(减去字符串结尾的零)后存储在常量 msg1Len 中。

注意在代码中使用函数序言和函数尾声。如上一章指出的，这些对于 GDB 正常运行是必需的。序言和尾声代码将在下一章进行说明。

下面使用 GDB 进行一些内存挖掘！输入以下内容：

```
gdb alive
```

然后在(gdb)提示符下输入以下内容：

```
disassemble main
```

输出如图 4-2 所示。

```
(gdb) disass main
Dump of assembler code for function main:
   0x00000000004004e0 <+0>:     push   rbp
   0x00000000004004e1 <+1>:     mov    rbp,rsp
   0x00000000004004e4 <+4>:     mov    eax,0x1
   0x00000000004004e9 <+9>:     mov    edi,0x1
   0x00000000004004ee <+14>:    movabs rsi,0x601030
   0x00000000004004f8 <+24>:    mov    edx,0xe
   0x00000000004004fd <+29>:    syscall
   0x00000000004004ff <+31>:    mov    eax,0x1
   0x0000000000400504 <+36>:    mov    edi,0x1
   0x0000000000400509 <+41>:    movabs rsi,0x60103f
   0x0000000000400513 <+51>:    mov    edx,0x13
   0x0000000000400518 <+56>:    syscall
   0x000000000040051a <+58>:    mov    rsp,rbp
   0x000000000040051d <+61>:    pop    rbp
   0x000000000040051e <+62>:    mov    eax,0x3c
   0x0000000000400523 <+67>:    mov    edi,0x0
   0x0000000000400528 <+72>:    syscall
   0x000000000040052a <+74>:    nop    WORD PTR [rax+rax*1+0x0]
End of assembler dump.
(gdb)
```

图 4-2　反汇编 alive

因此，在我们的计算机上，变量 msg1 似乎位于内存位置 0x601030；你可以使用以下方法进行查看：

```
x/s 0x601030
```

输出如图 4-3 所示。

```
(gdb) x/s 0x601030
0x601030 <msg1>:    "Hello, World!\n"
(gdb)
```

图 4-3　msg1 的内存位置

\n 代表 "new line(新行)"。在 GDB 中验证变量的另一种方法如下：

```
x/s &msg1
```

输出如图 4-4 所示。

```
(gdb) x/s &msg1
0x601030 <msg1>:        "Hello, World!\n"
(gdb)
```

图 4-4　用另一种方法验证变量

那么如何显示数值呢？

```
x/dw &radius
x/xw &radius
```

输出如图 4-5 所示。

```
(gdb) x/dw &radius
0x601053 <radius>:      357
(gdb) x/xw &radius
0x601053 <radius>:      0x00000165
(gdb)
```

图 4-5　数值

因此，你可以获得存储在内存位置的 radius 的十进制和十六进制值。对于浮点变量，请使用以下命令：

```
x/fg &pi
x/fx &pi
```

输出如图 4-6 所示。

```
(gdb) x/fg &pi
0x60105b <pi>:   3.1400000000000001
(gdb) x/fx &pi
0x60105b <pi>:   0x40091eb851eb851f
(gdb)
```

图 4-6　浮点数

注意到浮点错误了吗？

这里有一个你应该注意到的微妙之处。为便于演示，请打开生成的 alive.lst 文件。参见图 4-7。

第 4 章 ■ 你的下一个程序：Alive and Kicking

```
 1    1                                        ; alive.asm
 2    2                                            section .data
 3    3  00000000 48656C6C6F2C20576F-         msg1     db      "Hello, World!",10,0    ; string with NL and 0
 4    3  00000009 726C64210A00
 5    4                                        msg1Len  equ     $-msg1-1                ; measure the length, minus the 0
 6    5  0000000F 416C69766520616E64-         msg2     db      "Alive and Kicking!",10,0 ; string with NL and 0
 7    5  00000018 204B69636B696E6721-
 8    5  00000021 0A00
 9    6                                        msg2Len  equ     $-msg2-1                ; measure the length, minus the 0
10    7  00000023 6501000000000000            radius   dq      357                     ; no string, not displayable?
11    8  0000002B 1F85EB51B81E0940            pi       dq      3.14                    ; no string, not displayable?
12    9                                            section .bss
13   10                                            section .text
14   11                                                global main
15   12                                        main:
16   13  00000000 55                                   push    rbp                      ; function prologue
17   14  00000001 4889E5                               mov     rbp,rsp                  ; function prologue
18   15  00000004 B801000000                           mov     rax, 1                   ; 1 = write
19   16  00000009 BF01000000                           mov     rdi, 1                   ; 1 = to stdout
20   17  0000000E 48BE-                                mov     rsi, msg1                ; string to display
21   17  00000010 [0000000000000000]
22   18  00000018 BA0E000000                           mov     rdx, msg1Len             ; length of the string
23   19  0000001D 0F05                                 syscall                          ; display the string
24   20  0000001F B801000000                           mov     rax, 1                   ; 1 = write
25   21  00000024 BF01000000                           mov     rdi, 1                   ; 1 = to stdout
26   22  00000029 48BE-                                mov     rsi, msg2                ; string to display
27   22  0000002B [0F00000000000000]
28   23  00000033 BA13000000                           mov     rdx, msg2Len             ; length of the string
29   24  00000038 0F05                                 syscall                          ; display the string
30   25  0000003A 4889EC                               mov     rsp,rbp                  ; function epilogue
31   26  0000003D 5D                                   pop     rbp                      ; function epilogue
32   27  0000003E B83C000000                           mov     rax, 60                  ; 60 = exit
33   28  00000043 BF00000000                           mov     rdi, 0                   ; 0 = success exit code
34   29  00000048 0F05                                 syscall                          ; quit
```

图 4-7　alive.lst

看一下第 10 和 11 行，在左边可以找到 radius 和 pi 的十六进制表示形式。不是 0165，而是 6501；不是 40091EB851EB851F，而是 1F85EB51B81E0940。因此，字节(1 字节是两个十六进制数字)的顺序是反的！

这种特性称为字节序。big-endian(大端)格式存储数字的方式与我们惯用的方式相同，最高有效数字从左开始。little-endian(小端)格式从左开始存储最低有效数字。英特尔处理器使用 little-endian(小端)字节序，这在查看十六进制代码时会非常混乱。

为什么他们有这么奇怪的名字，像大端(big-endian)和小端(little-endian)？

1726 年，乔纳森·斯威夫特(Jonathan Swift)写了一本著名的小说《格列佛游记》。小说中出现了两个虚构的岛屿，利里普特岛和布莱夫斯库。利里普特的居民正与布莱夫斯库的人们就如何打破鸡蛋而争辩：在较小的一端还是较大的一端。利里普特居民认为是小端(little-endian)，喜欢从较小的一端打破鸡蛋。布莱夫斯库居民认为是大端(big-endian)。现在你应该发现，现代计算的传统植根于遥远的过去！

花点时间单步执行程序(break main、run、next 等)。你可看到 GDB 跨越了函数序言。编辑源代码，删除函数序言和尾声，然后重新对程序执行 make 操作。使用 GDB 再次单步执行。在我们的例子中，GDB 拒绝单步执行，而是完全执行程序。在使用另一个基于 NASM 的汇编器 YASM 进行汇编时，我们可以安全地省略序言和尾声代码，并使用 GDB 逐步完成代码。有时有必要进行实验、修补和 Google 搜索！

4.2 打印

alive 程序打印两个字符串：

Hello, World!
Alive and Kicking!

但是，还有两个变量没有定义为字符串：radius 和 pi。打印这些变量比打印字符串要复杂一些。要以与 msg1 和 msg2 类似的方式打印这些变量，必须将 radius 和 pi 值转换为字符串。将转换代码添加到程序中是完全可行的，但是这会使这个小程序在此时变得过于复杂，因此我们将用一点小手段。我们将从 C 语言借用一个通用函数 printf 并将其包含在程序中。如果这让你心烦，请耐心等待。当你成为更高级的汇编程序程序员时，可以编写自己的函数来转换/打印数字。

为了在汇编程序中引入 printf，我们将从一个简单的程序开始。修改第一个程序 hello.asm，如代码清单 4-3 所示。

代码清单 4-3：修改第一个程序 hello.asm

```
; hello4.asm
extern      printf        ; 声明外部函数
section .data
    msg     db "Hello, World!",0
    fmtstr  db "This is our string: %s",10,0  ; 打印格式
section .bss
section .text
    global main
main:
    push    rbp
    mov     rbp,rsp
    mov     rdi, fmtstr    ; printf 的第一个参数
    mov     rsi, msg       ; printf 的第二个参数
    mov     rax, 0         ; 不涉及 xmm 寄存器
    call    printf         ; 调用函数
    mov     rsp,rbp
    pop     rbp
    mov     rax, 60        ; 60 表示退出
    mov     rdi, 0         ; 0 表示成功退出代码
    syscall                ; 退出
```

因此，我们首先告诉汇编器(和链接器)，将使用一个称为 printf 的外部函数。我们创建了一个用于格式化的字符串，来确定 printf 将如何显示 msg。格式字符串的

语法类似于 C 中的语法。如果你有使用 C 的经验，那么肯定会识别格式字符串。%s 是要打印的字符串的占位符。

不要忘记函数序言和尾声。将 msg 的地址放入 rsi，并将 fmtstr 的地址放入 rdi。清除 rax，在这种情况下，这意味着 xmm 寄存器中没有要打印的浮点数。浮点数和 xmm 寄存器将在第 11 章中介绍。

代码清单 4-4 展示了 makefile。

代码清单 4-4：hello4.asm 的 makefile

```
#hello4.asm 的 makefile
hello4: hello4.o
    gcc -o hello4 hello4.o -no-pie
hello4.o: hello4.asm
    nasm -f elf64 -g -F dwarf hello4.asm -l hello4.lst
```

确保在 makefile 中添加了 -no-pie 标志；否则，使用 printf 将导致错误。请记住，从第 1 章开始，当前的 gcc 编译器会生成与位置无关的可执行(pie)代码，以使其更安全。结果之一是我们不能再简单地使用外部函数。为了避免这种复杂性，我们使用标志-no-pie。

编译并运行程序。可以在互联网上搜索 C printf 函数来了解更多可能的格式。如你所见，借助 printf，我们可以灵活地将输出格式化为打印整数、浮点数、字符串、十六进制数据等。printf 函数要求字符串以 0(NULL)结尾。如果省略 0，则 printf 会一直显示所有内容，直至找到 0 为止。汇编中并不需要以 0 结尾的字符串，但是对于 printf、GDB 和某些 SIMD 指令，这是必需的(SIMD 将在第 26 章中介绍)。

输出如图 4-8 所示。

```
jo@UbuntuDesktop:~/Desktop/linux64/gcc/05  hello4$ make
nasm -f elf64 -g -F dwarf hello4.asm -l hello4.lst
gcc -o hello4 hello4.o
jo@UbuntuDesktop:~/Desktop/linux64/gcc/05  hello4$ ./hello4
This is our string: Hello, World!
jo@UbuntuDesktop:~/Desktop/linux64/gcc/05  hello4$
```

图 4-8 alive.lst

回到我们的程序！我们现在可以使用 printf 打印变量 radius 和 pi。代码清单 4-5 展示了源代码。现在，你知道该怎么做：创建源代码，复制或创建/修改 makefile，然后就可以了。

代码清单 4-5：alive2.asm 的 makefile

```
; alive2.asm
section .data
    msg1    db      "Hello, World!",0
```

```
        msg2     db      "Alive and Kicking!",0
        radius   dd      357
        pi       dq      3.14
        fmtstr   db      "%s",10,0       ;用于打印一个字符串的格式
        fmtflt   db      "%lf",10,0      ;用于打印一个浮点数的格式
        fmtint   db      "%d",10,0       ;用于打印一个整数的格式
section .bss
section .text
extern      printf
    global main
main:
    push rbp
    mov rbp,rsp
; print msg1
    mov    rax, 0            ; 没有浮点数
    mov rdi, fmtstr
    mov rsi, msg1
    call printf
; print msg2
    mov    rax, 0            ; 没有浮点数
    mov rdi, fmtstr
    mov rsi, msg2
    call printf
; print radius
    mov    rax, 0            ; 没有浮点数
    mov rdi, fmtint
    mov rsi, [radius]
    call printf
; print pi
    mov    rax, 1            ; 使用一个 xmm 寄存器
    movq   xmm0, [pi]
    mov    rdi, fmtflt
    call   printf

    mov    rsp,rbp
    pop    rbp
ret
```

我们添加了三个用于格式化打印输出的字符串。将格式字符串放入 rdi，将 rsi 指向要打印的项目，将 0 放入 rax 以指示不涉及浮点数，然后调用 printf。要打印浮点数，请使用 movq 指令移动要在 xmm0 中显示的浮点值。我们使用一个 xmm 寄存器，所以将 1 放入 rax 中。在后续章节中，我们将更多地讨论用于浮点计算的 xmm

寄存器和 SIMD 指令。

请注意将 radius 和 pi 括起来的方括号[]。

```
mov rsi, [radius]
```

这意味着获取 radius 的内容并放入 rsi 中。字符串函数 printf 需要一个内存地址，但对于数字，它期望一个值，而不是内存地址。记住这一点。

程序的退出有点不同，不是我们所熟悉的代码：

```
mov rax, 60         ; 60 = exit
mov rdi, 0          ; 0 = success exit code
syscall             ; quit
```

我们使用等效的方式：

```
ret
```

注意，printf 采用格式字符串，该格式字符串可以采用不同的形式，并且可以转换打印值的性质(整数、双精度、浮点数等)。但有时这种转换是无意的，可能会令人困惑。如果确实想知道程序中的寄存器或变量的值(内存位置)，请使用调试器并检查寄存器或内存位置。

图 4-9 展示了 alive2 程序的输出。

```
jo@UbuntuDesktop:~/Desktop/linux64/gcc/06 alive2$ make
nasm -f elf64 -g -F dwarf alive2.asm -l alive2.lst
gcc -o alive2 alive2.o -no-pie
jo@UbuntuDesktop:~/Desktop/linux64/gcc/06 alive2$ ./alive2
Hello, World!
Alive and Kicking!
357
3.140000
jo@UbuntuDesktop:~/Desktop/linux64/gcc/06 alive2$
```

图 4-9　alive2 的输出

4.3　小结

本章内容：
- GDB 的其他功能
- 函数序言和尾声
- 大端与小端
- 使用 C 语言函数 printf 打印字符串、整数和浮点数

第 5 章

汇编是基于逻辑的

是时候学习一些逻辑理论了。不必惊慌,因为我们只学习需要的内容:NOT、OR、XOR 和 AND。

在本章中,0 表示错误(false),1 表示正确(true)。

5.1 NOT

可参考表 5-1。

表 5-1 NOT

A	0	1
NOT A	1	0

将每个 0 转换为 1,并将每个 1 转换为 0。

下面看一个示例:

```
A     =   11001011
NOT A =   00110100
```

5.2 OR

可参考表 5-2。

表 5-2 OR

A	0	1	0	1
B	0	0	1	1
A OR B	0	1	1	1

如果 A 或 B 或两者都为 1，则结果为 1。

下面看一个示例：

```
A       =   11001011
B       =   00011000
A OR B  =   11011011
```

5.3 XOR

可参考表 5-3。

表 5-3 XOR

A	0	1	0	1
B	0	0	1	1
A XOR B	0	1	1	0

异或：如果 A 或 B 的值为 1(而非均为 1)，则结果为 1。如果 A 和 B 均为 1 或 0，则结果为 0。

下面看一个示例：

```
A        =   11001011
B        =   00011000
A XOR B  =   11010011
```

XOR 可以用作清除寄存器的汇编指令。

```
A        =   11001011
A        =   11001011
A XOR A  =   00000000
```

因此，xor rax,rax 等于 mov rax,0。但是 xor 的执行速度比 mov 快。也可以使用 xor 修改浮点数的符号。下面是一个 32 位浮点数示例：

```
A  = 17.0 = 0x41880000 = 01000001 10001000 00000000 00000000
B  = -0.0 = 0x80000000 = 10000000 00000000 00000000 00000000
```

```
A XOR B =-17.0 = 0xC1880000 = 11000001 10001000 00000000 00000000
```

使用 www.binaryconvert.com/result_float.html 上的工具进行验证。

请注意，如果要更改整数的符号，请使用 0 去减或使用 neg 指令。

5.4　AND

可参考表 5-4。

表 5-4　AND

A	0	1	0	1
B	0	0	1	1
A AND B	0	0	0	1

如果 A 和 B 的值都为 1，则结果为 1，否则为 0。

下面看一个示例：

```
A =         11001011
B =         00011000
A AND B =   00001000
```

AND 指令可用作选择和检查位的掩码。

在本例中，B 用作掩码以从 A 中选择位 3 和 6(最低、最右边的位的索引为 0)：

```
A =         11000011
B =         01001000
A AND B =   01000000
```

在本例中，第 6 位被置位，而第 3 位未置位。

AND 指令还可用于舍入数字，并且在 16 字节边界上舍入地址特别有用。稍后将使用它对齐堆栈。

16 和 16 的倍数都以二进制 0 或 0000 结尾。

```
address = 0x42444213 = 01000010010001000100001000010011
mask    = 0xfffffff0 = 11111111111111111111111111110000
rounded = 0x42444210 = 01000010010001000100001000010000
```

这里将地址的最低字节取整。如果地址已经以零字节结尾，AND 指令将不会更改任何内容。验证舍入后的地址是否可被 16 整除。使用联机工具进行转换(例如，www.binaryconvert.com/convert_unsigned_int.html)。

5.5 小结

本章内容：
- 逻辑运算符
- 如何将逻辑运算符用作汇编指令

第 6 章

数据显示调试器

数据显示调试器(DDD)是带有图形用户界面的 Linux 调试工具。请现在安装它(使用 sudo apt install ddd),因为我们将在本章后面使用它。我们在本章中编写的程序将没有输出。我们将研究代码执行并将内容注册到 DDD。

6.1 使用 DDD

代码清单 6-1 展示了示例代码。

代码清单 6-1: move.asm

```
; move.asm
section .data
      bNum    db 123
      wNum    dw 12345
      dNum    dd 1234567890
      qNum1   dq 1234567890123456789
      qNum2   dq 123456
      qNum3   dq 3.14
section .bss
section .text
      global main
main:
push rbp
mov rbp,rsp
      mov rax, -1          ; 把-1 放入 rax 中,即 rax 中的所有位都是 1
      mov al, byte [bNum]  ; 不清除 rax 的高位(upper bits)
```

```
        xor rax,rax                 ; 清除 rax
        mov al, byte [bNum]         ; 现在 rax 中是正确的值

        mov rax, -1                 ; 把-1 放入 rax 中，即 rax 中的所有位都是 1
        mov ax, word [wNum]         ; 不清除 rax 的高位(upper bits)
        xor rax,rax                 ; 清除
        mov ax, word [wNum]         ; 现在 rax 中是正确的值

        mov rax, -1                 ; 把-1 放入 rax 中，即 rax 中的所有位都是 1
        mov eax, dword [dNum]       ; 不清除 rax 的高位(upper bits)

        mov rax, -1                 ; 把-1 放入 rax 中，即 rax 中的所有位都是 1
        mov rax, qword [qNum1]      ; 不清除 rax 的高位(upper bits)
        mov qword [qNum2], rax      ; 有一个操作数永远是一个寄存器
        mov rax, 123456             ; 源操作数是一个即时数

        movq xmm0, [qNum3]          ;浮点数指令

mov rsp,rbp
pop rbp

ret
```

将源文件另存为 move.asm，然后编译并运行它以查看是否正常工作。运行它时不应显示任何内容。在命令行提示符下输入以下内容：

```
ddd move
```

你将看到一个布局相当过时的 GUI(参见图 6-1)。DDD 是一个古老的开源工具，显然没有人愿意将它改编成适应我们今天使用的 GUI 标准。

有一个显示源代码的窗口，以及一个可以输入 GDB 命令的窗口。还有一个浮动面板，你可在其中单击 Run、Step、Stepi 等。单击菜单中的 Source，然后选择显示行号。在同一菜单中，你可以选择一个包含汇编代码的窗口。

将光标放在 "main:" 前面，右击并选择 Break，或选择顶部菜单中的 Stop 图标。在浮动面板上单击 Run 开始调试。单击顶部菜单栏中的 Status，然后选择 Registers。单击 Step 执行指令。现在，你可以了解逐步执行该程序时寄存器是如何更改的了。如果要检查诸如 qNum1 或 bNum 的内存地址，则可以使用顶部的 Data 菜单项。首先转到 View，使数据窗口可见。然后单击 Data 菜单项中的 Memory。有关如何调查内存的示例，参见图 6-2。由于 DDD 的界面比较复杂，因此使用 GDB 输入窗口有时比使用菜单快得多。

第 6 章 ■ 数据显示调试器

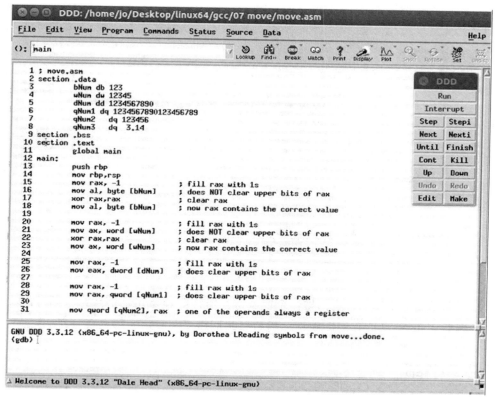

图 6-1　DDD 的主界面

　　DDD 建立在 GDB 之上，因此我们需要使用函数序言和尾声，以避免出现问题。请注意，单步执行程序时，DDD 会忽略序言。

　　该代码的目的是向你展示使用 mov 命令时寄存器的内容会发生什么变化。在 DDD 中打开 Registers 窗口(选择 Status | Registers)。注意，最初的 rax 包含-1；这意味着 rax 中的所有位均为 1。如果你不了解原因，请回到有关二进制的章节。你将看到，如果将数字移到 al 或 ax 中，则 rax 中的高位不会清除为 0，因此 rax 寄存器中的值与 al 或 ax 的值不同。在我们的示例中，如果 rax 包含 0xffffffffffffff7b，则该值为大的负数。但正如我们所期望的，al 包含 0x7b，十进制为 123。这可能不是你的意图。如果你在计算中错误地使用了 rax 而不是 al，那么结果将是错误的！但是，随着继续单步执行代码，你将看到将 32 位值移至 64 位寄存器时，将清除 64 位寄存器中的高位。当你将一个值移入 eax 时，rax 的高位将被清除。这一点很重要！

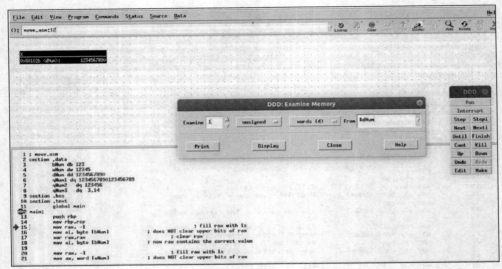

图 6-2　使用 DDD 进行内存调查

下面总结这个练习，我们将一个值从寄存器移至 qNum2。请注意方括号，以告诉汇编器 qNum2 是内存中的地址。最后，我们将"即时值"放入寄存器。

6.2　小结

本章内容：

- DDD 尽管已过时，但仍可用作调试器，它是基于 GDB 的。
- 将值复制到 8 位或 16 位寄存器中不会清除 64 位寄存器的高位部分。
- 但是，将值复制到 32 位寄存器中会清除 64 位寄存器的高位部分。

第 7 章

跳转和循环

相信你也会认为诸如 DDD 的可视调试器非常有用，尤其是在研究大型程序时。在本章中，我们将介绍 SASM(SimpleASM)。它是一个开放源代码、跨平台的集成开发环境(IDE)。它具有语法突出显示和图形调试功能。对于汇编程序员来说，这是一个不错的工具！

7.1 安装 SimpleASM

转到 https://dman95.github.io/SASM/english.html，为操作系统选择版本，然后安装它。对于 Ubuntu 18.04，进入目录 xUbuntu_18.04/amd64/并使用以下命令下载和安装 sasm_3.10.1_amd64.deb 软件包：

```
sudo dpkg -i sasm_3.10.1_amd64.deb
```

如果收到有关依赖关系问题的错误消息，请安装缺少的软件包，然后重试。也可以尝试以下方法：

```
sudo apt --fix-broken install
```

通常，这将安装所有必需的软件包。

7.2 使用 SASM

通过在命令行输入 sasm 来启动 SASM，然后选择语言。SASM 启动后，如果报错，例如无法加载模块 canberra-gtk-module，请安装以下软件包：

```
sudo apt install libcanberra-gtk*
```

这将安装一堆文件，之后就不会再看到报错了。

在 SASM 中，进入 Settings 对话框，如图 7-1 所示。在 Common 选项卡上，为 Show all registers in debug 选择 Yes。

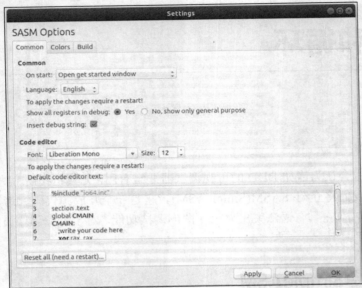

图 7-1　SASM 设置对话框——Common 选项卡

在 Build 选项卡中修改设置，如图 7-2 所示。

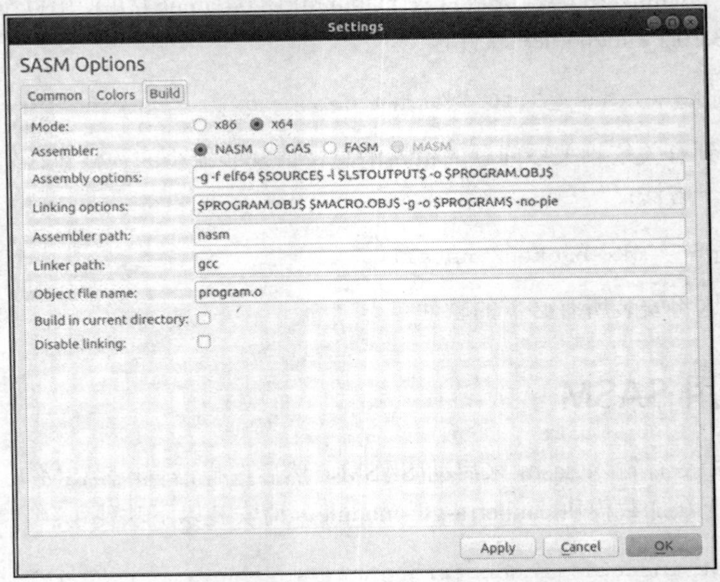

图 7-2　SASM 设置对话框——Build 选项卡

这里要非常小心，因为设置必须完全如图所示。如果空格太多，甚至隐藏在一行的末尾处，SASM 将无法执行你想要的操作。准备就绪后，单击 OK 按钮，然后重新启动 SASM。

使用 SASM 启动新项目时，会在编辑器窗口中看到一些默认代码。我们不会使用该代码，因此你可以删除它。在命令行输入以下内容：

```
sasm jump.asm
```

如果 jump.asm 不存在，SASM 将启动一个新的编辑器窗口，然后删除默认代码。如果文件已存在，它将在编辑器窗口中打开。

代码清单 7-1 展示了 jump.asm 的代码。

代码清单 7-1：jump.asm

```
; jump.asm
extern printf
section .data
     number1 dq 42
     number2 dq 41
     fmt1 db "NUMBER1 > = NUMBER2",10,0
     fmt2 db "NUMBER1 < NUMBER2",10,0

section .bss
section .text
     global main
main:
     push rbp
     mov  rbp,rsp
     mov  rax, [number1]   ; 把数字放入寄存器
     mov  rbx, [number2]
     cmp  rax,rbx          ; 比较 rax 和 rbx
     jge  greater          ; 如果 rax 大于或等于 rbx，则转到 greater
mov rdi,fmt2               ; 如果 rax 小于 rbx，则从这里继续
mov rax,0                  ; 不涉及 xmm 寄存器
     call printf           ; 显示 fmt2
     jmp  exit             ; 跳转到 exit
greater:
     mov  rdi,fmt1         ; 当 rax 大于或等于 rbx
     mov  rax,0            ; 不涉及 xmm 寄存器
     call printf           ; 显示 fmt1
exit:
     mov  rsp,rbp
```

```
        pop   rbp
        ret
```

将代码复制到 SASM 编辑器窗口中；默认情况下，SASM 将使用语法高亮显示。输入完成后，单击顶部的绿色三角形图标(意思是"运行")。如果一切正常，将在 Output 区域中看到输出，如图 7-3 所示。

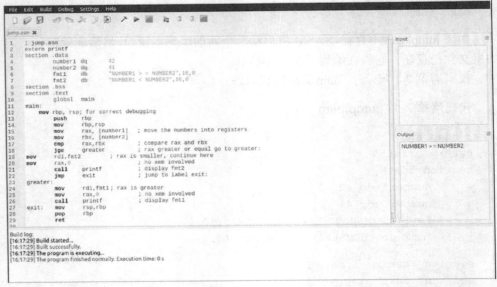

图 7-3　SASM 输出

在 SASM 中保存文件时，将保存源代码。如果要保存可执行文件，则需要在 File 菜单中选择 Save.exe。

要开始调试，请在 main:标签左侧单击编号的左边距。这将在 main:标签及其行号之间放置一个红色圆圈，这就是一个断点。然后在顶部菜单上单击带有瓢虫标记的绿色三角形。在顶部菜单中，选择 Debug，然后选择 Show Registers 和 Show Memory。屏幕上还会显示许多其他窗口：Registers、Memory 和 GDB 命令行小程序。

使用 Step 图标，可以遍历代码并查看寄存器值是如何变化的。要研究变量如何变化，请右击 section .data 中的变量声明，然后选择 Watch。变量将被添加到内存窗口中，SASM 将尝试猜测类型。如果 SASM 显示的值与预期不符，请手动将类型更改为正确的格式。使用 SASM 进行调试时，会添加以下代码行进行正确的调试：

```
mov rbp, rsp; for correct debugging
```

这一行可能会使其他调试器(如 GDB)感到困惑，因此在从命令行单独运行 GDB

之前，请确保将其从代码中删除。

在 SASM 菜单的 Settings | Common 中，确保将 Show all registers in debug 选择为 Yes。在 SASM 中进行调试时，在寄存器(register)窗口中向下滚动。你将在底部看到 16 个 ymm 寄存器，每个寄存器在括号之间有两个值。第一个值是对应的 xmm 寄存器。在讨论 SIMD 时，我们将更详细地解释这些寄存器。

顺便说一句，图 7-4 展示了编译和运行程序后的输出。

```
jo@UbuntuDesktop:~/Desktop/linux64/gcc/08 jump$ make
nasm -f elf64 -g -F dwarf jump.asm -l jump.lst
gcc -o jump jump.o
jo@UbuntuDesktop:~/Desktop/linux64/gcc/08 jump$ ./jump
NUMBER1 > = NUMBER2
jo@UbuntuDesktop:~/Desktop/linux64/gcc/08 jump$
```

图 7-4　jump.asm 的输出

在这个程序中，我们使用了一条比较指令 cmp 和两条跳转指令 jge 和 jmp。cmp 在这里用来比较两个操作数，在本例中是两个寄存器。两个操作数之一也可以是内存操作数，第二个操作数可以是即时值。在任何情况下，两个操作数的大小必须相同(字节、字等)。cmp 指令将设置或清除标志寄存器中的标志。

标志是位于 rflags 寄存器中的位，可以设置为 1 或清除为 0，具体取决于许多条件。在我们的例子中，重要的是零(zero)标志(ZF)、溢出(overflow)标志(OF)和符号(sign)标志(SF)。可以使用调试器查看这些标志。使用 SASM 可以轻松查看所有寄存器(包括标志寄存器，在 SASM 中被称为 eflags)的内容。cmp 操作数中的不同值将导致设置或清除不同的标志。对这些值进行一些实验，以查看这些标志发生了什么。

如果要使用这些标志，必须在 cmp 指令之后立即计算它们。如果在计算 rflags 之前执行其他指令，则标志可能已更改。在此处的程序中，用 jge 计算标志，意思是"大于或等于时进行跳转"。如果条件满足，则执行跳转到 jge 指令之后的标签。如果不满足条件，则继续执行 jge 之后的指令。表 7-1 列出了一些常见情况，你可以在英特尔手册中查找更多信息。

表 7-1　跳转指令和标志

指令	标志	含义	应用
je	ZF=1	如果相等则跳转	有符号、无符号
jne	ZF=0	如果不相等则跳转	有符号、无符号
jg	((SF XOR OF) OR ZF) = 0	如果大于则跳转	有符号
jge	(SF XOR OF) = 0	如果大于或等于则跳转	有符号
jl	(SF XOR OF) = 1	如果低于则跳转	有符号

(续表)

指令	标志	含义	应用
jle	((SF XOR OF) OR ZF) = 1	如果低于或等于则跳转	有符号
ja	(CF OR ZF) = 0	如果高于则跳转	无符号
jae	CF=0	如果高于或等于则跳转	无符号
jb	CF=1	如果少于则跳转	无符号
jbe	(CF OR ZF) = 1	如果少于或等于则跳转	无符号

在程序中，我们还有一个无条件跳转指令 jmp。如果程序执行命中该指令，则不管标志或条件如何，程序都会跳转到 jmp 后面指定的标签。

一种更复杂的跳转形式是循环，这意味着重复一组指令，直到满足(或不满足)一个条件。代码清单 7-2 展示了一个示例。

代码清单 7-2：jumploop.asm

```
; jumploop.asm
extern printf
section .data
    number dq 5
    fmt   db "The sum from 0 to %ld is %ld",10,0
section .bss
section .text
    global main
main:
    push rbp
    mov rbp, rsp
    mov rbx,0      ; 计数器
    mov rax,0      ; 和(sum)将存放在 rax 中
jloop:
    add rax, rbx
    inc rbx
    cmp rbx,[number]  ; 比较 rbx 和 number 是否相等
    jle jloop         ;如果 rbx 小于或等于 number，继续循环
                      ;否则，继续往下执行
    mov rdi,fmt       ; 准备进行显示
    mov rsi, [number]
    mov rdx,rax
    mov rax,0
    call printf
    mov rsp,rbp
```

```
        pop     rbp
        ret
```

程序将 0~number 的值之间的所有数字相加。使用 rbx 作为计数器，使用 rax 跟踪总和。我们创建了一个循环，它是 jloop:和 jle jloop 之间的代码。在循环中，我们将 rbx 中的值加到 rax 中，将 rbx 增加 1，然后比较 rbx 是否等于 number。如果 rbx 中的值小于或等于 number，则重新启动循环；否则，继续执行循环后的指令并准备打印结果。我们使用算术指令 inc 来增加 rbx。将在后续章节中讨论算术指令。

代码清单 7-3 展示了另一种编写循环的方法。

代码清单 7-3：betterloop.asm

```
; betterloop
extern printf
section .data
    number  dq  5
    fmt     db  "The sum from 0 to %ld is %ld",10,0
section .bss
section .text
    global main
main:
    push    rbp
    mov     rbp,rsp
    mov     rcx,[number]    ; 使用 number 初始化 rcx
    mov     rax, 0
bloop:
    add     rax,rcx         ; 把 rcx 添加到和(sum)
    loop    bloop           ; 每次循环 rcx 都递减 1，直到等于 0
    mov     rdi,fmt         ; 如果 rcx = 0，则继续往下执行
    mov     rsi, [number]   ; 要显示的和(sum)
    mov     rdx, rax
    mov     rax,0           ; 没有浮点数
    call    printf          ; 显示和(sum)
    mov     rsp,rbp
    pop     rbp
    ret
```

这里有一个特殊的循环指令，它使用 rcx 作为递减循环计数器。每次循环执行时，rcx 都会自动减 1，只要 rcx 不等于 0，循环就会再次执行。这种做法输入的代码更少。

一个有趣的实验是将 1000000000(一个 1 和 9 个 0)放入 number，然后重新编译

并运行前面的两个程序。可以使用 Linux 的 time 命令对速度进行计时，如下所示：

```
time ./jumploop
time ./betterloop
```

注意，betterloop 比 jumploop 慢(见图 7-5)！使用循环指令很方便，但要以执行性能为代价。我们使用 Linux 时间指令来衡量性能。稍后将展示更合适的方法来研究和调优程序代码。

```
jo@UbuntuDesktop:~/Desktop/linux64/gcc/09 loopcompare$ time ./betterloop_long
The sum from 0 to 1000000000 is 500000000500000000

real    0m1.731s
user    0m1.726s
sys     0m0.004s
jo@UbuntuDesktop:~/Desktop/linux64/gcc/09 loopcompare$ time ./jumploop_long
The sum from 0 to 1000000000 is 500000000500000000

real    0m0.404s
user    0m0.391s
sys     0m0.008s
jo@UbuntuDesktop:~/Desktop/linux64/gcc/09 loopcompare$
```

图 7-5　比较循环和跳转

你可能想知道，既然有 SASM 这样的工具，为什么还要费心使用 DDD。你将意识到在 SASM 中不能查看堆栈，但使用 DDD 却可以。稍后将再次使用 DDD。

7.3　小结

本章内容：
- 如何使用 SASM
- 如何使用跳转指令
- 如何使用 cmp 指令
- 如何使用循环指令
- 如何计算标志

第 8 章

内存

内存被处理器用作数据和指令的存储空间。我们已经讨论过寄存器,它是高速访问存储的地方。访问内存比访问寄存器慢得多。但寄存器的数量是有限的。内存大小的理论限制是 2^{64} 个地址,即 18 446 744 073 709 551 616 个。由于实际的设计问题,你不能使用那么多内存!现在是时候更详细地研究内存了。

8.1 探索内存

代码清单 8-1 展示了我们在讨论内存时将使用的示例。

代码清单 8-1:memory.asm

```
; memory.asm
section .data
    bNum    db  123
    wNum    dw  12345
    warray  times   5 dw 0      ; 包含 5 个元素的数组
                                ; 元素都是 0
    dNum    dd  12345
    qNum1   dq  12345
    text1   db  "abc",0
    qNum2   dq  3.141592654
    text2   db  "cde",0
section .bss
    bvar    resb 1
    dvar    resd 1
```

```
        wvar resw 10
        qvar resq 3
section .text
        global main
main:
        push    rbp
        mov     rbp, rsp
        lea     rax, [bNum]
        mov     rax, bNum           ;在 rax 中加载 bNum 的地址
        mov     rax, [bNum]         ;在 rax 中加载 bNum 的值
        mov     [bvar], rax         ;将 rax 中的数据加载到 bvar 的地址
        lea     rax, [bvar]         ;在 rax 中加载 bvar 的地址
        lea     rax, [wNum]         ;在 rax 中加载 wNum 的地址
        mov     rax, [wNum]         ;在 rax 中加载 wNum 的内容
        lea     rax, [text1]
        mov     rax, text1          ;在 rax 中加载 text1 的地址
        mov     rax, text1+1
        lea     rax, [text1+1]      ;在 rax 中加载第二个字符
        mov     rax, [text1]        ;在 rax 中从 text1 开始加载
        mov     rax, [text1+1]      ;在 rax 中从 text1+1 开始加载
        mov     rsp,rbp
        pop     rbp
        ret
```

编译这个程序，它没有输出。使用调试器逐步执行每条指令。SASM 在这里很有帮助。

我们定义了一些大小不同的变量，包括五个用零填充的双字数组。我们还在.bss 段中定义了一些变量。在调试器中查看 rsp，即栈指针；它有非常高的价值。栈指针指向内存中的高位地址。堆栈是内存中用于临时存储数据的区域。堆栈将随着存储数据的增加而增长，并且堆栈将从高地址到低地址向下增长。每次将数据放入堆栈时，栈指针 rsp 都会变小。我们将在单独的章节中讨论堆栈，但请记住，堆栈是高位内存中的一个位置。见图 8-1。

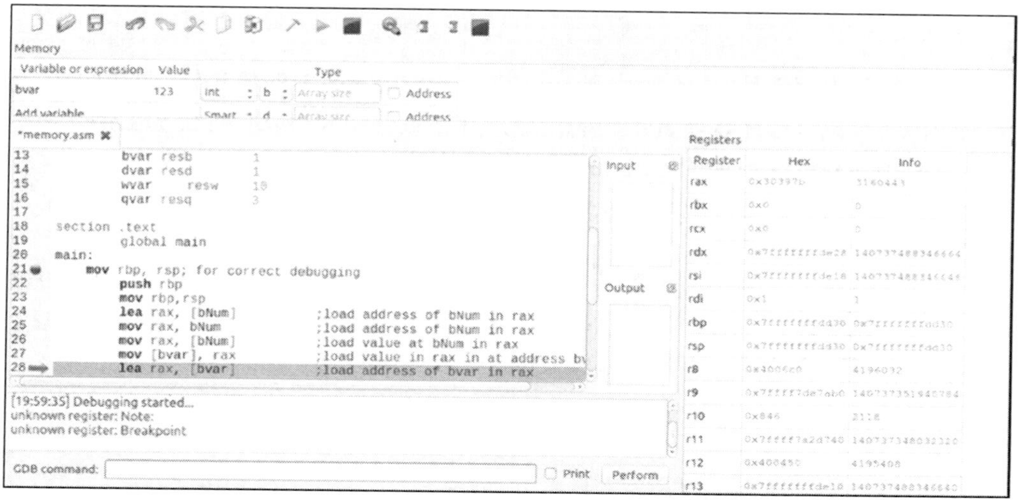

图 8-1　rsp 包含高位内存中的一个地址

我们使用 lea(意思是"加载有效地址")指令将 bNum 的内存地址加载到 rax 中。使用 mov 可获得相同结果，bNum 周围没有方括号。如果在 mov 指令中使用方括号[]，则会将值(而不是 bNum 的地址)加载到 rax 中。但是我们不只是将 bNum 加载到 rax 中。由于 rax 是 64 位(或 8 字节)寄存器，因此更多字节被加载到 rax 中。我们的 bNum 是 rax(小端)中最右边的字节；这里，仅对寄存器 al 感兴趣。当你要求 rax 仅包含值 123 时，首先必须清除 rax，如下所示：

```
xor rax, rax
```

使用：

```
mov al, [bNum]
```

替代：

```
mov rax, [bNum]
```

注意要在内存中移动的数据的大小。例如，请看以下内容：

```
mov [bvar],rax
```

使用这条指令，将 rax 中的 8 个字节移到地址 bvar。如果仅打算将 123 放入 bvar，则可以使用调试器检查是否覆盖了内存中的另外 7 个字节(在 SASM 内存窗口中为 bvar 选择 d 类型)！这可能会在程序中引入讨厌的 bug。为避免这种情况，请将指令替换为以下内容：

```
mov [bvar],al
```

将内容从内存地址 text1 加载到 rax 中时,请注意 rax 中的值是如何使用小端表示法的。逐步执行程序以研究不同的指令,并更改值和大小以查看会发生什么。

有两种加载内存地址的方法: mov 和 lea。使用 lea 可以使你的代码更具可读性,因为每个人都可以立即看到你正在这里处理地址。也可以使用 lea 加快计算速度,但这里不会使用 lea。

启动 gdb 内存,然后反汇编 main 并查看带有内存地址的左列(见图 8-2)。正如在上一章中所解释的那样,不要忘记首先删除 SASM 添加的行以进行正确的调试。在本例中,第一条指令位于地址 0x4004a0。

```
(gdb) disass main
Dump of assembler code for function main:
   0x00000000004004a0 <+0>:     mov    rbp,rsp
   0x00000000004004a3 <+3>:     push   rbp
   0x00000000004004a4 <+4>:     mov    rbp,rsp
   0x00000000004004a7 <+7>:     lea    rax,ds:0x601028
   0x00000000004004af <+15>:    movabs rax,0x601028
   0x00000000004004b9 <+25>:    mov    rax,QWORD PTR ds:0x601028
   0x00000000004004c1 <+33>:    mov    QWORD PTR ds:0x601058,rax
   0x00000000004004c9 <+41>:    lea    rax,ds:0x601058
   0x00000000004004d1 <+49>:    lea    rax,ds:0x601029
   0x00000000004004d9 <+57>:    mov    rax,QWORD PTR ds:0x601029
   0x00000000004004e1 <+65>:    lea    rax,ds:0x601041
   0x00000000004004e9 <+73>:    movabs rax,0x601041
   0x00000000004004f3 <+83>:    movabs rax,0x601042
   0x00000000004004fd <+93>:    lea    rax,ds:0x601042
   0x0000000000400505 <+101>:   mov    rax,QWORD PTR ds:0x601041
   0x000000000040050d <+109>:   mov    rax,QWORD PTR ds:0x601042
   0x0000000000400515 <+117>:   mov    rsp,rbp
   0x0000000000400518 <+120>:   pop    rbp
   0x0000000000400519 <+121>:   ret
   0x000000000040051a <+122>:   nop    WORD PTR [rax+rax*1+0x0]
End of assembler dump.
(gdb)
```

图 8-2 用 GDB 对 main 进行反汇编

现在,将在命令行中使用 readelf。请记住,我们要求 NASM 使用 ELF 格式进行汇编(请参阅 makefile)。readelf 是一个命令行工具,用于获取有关可执行文件的更多信息。如果想进一步了解链接器,那么这里有一个有趣的信息源:

Linkers and Loaders, John R. Levine, 1999,"软件工程和编程"中的 Morgan Kaufmann 系列。

这里有 ELF 格式的简短论述:

https://linux-audit.com/elf-binaries-on-linuxunderstanding-and-analysis/

第 8 章 ■ 内存

或

`https://www.cirosantilli.com/elf-hello-world/`

相信你可能已经猜到了，还可以在命令行输入以下内容来了解 elf：

`man elf`

出于我们的目的，在命令行输入以下内容：

`readelf --file-header ./memory`

你将获得有关可执行内存的一些常规信息。查看入口点地址 0x4003b0，这是程序入口的内存位置。因此，在程序入口和代码开头之间存在一些开销，如 GDB(0x4004a0)所示。该标头提供了有关操作系统和可执行代码的其他信息。请参见图 8-3。

```
jo@ubuntu18:~/Desktop/linux64/gcc/10 memory$ readelf --file-header ./memory
ELF Header:
  Magic:   7f 45 4c 46 02 01 01 00 00 00 00 00 00 00 00 00
  Class:                             ELF64
  Data:                              2's complement, little endian
  Version:                           1 (current)
  OS/ABI:                            UNIX - System V
  ABI Version:                       0
  Type:                              EXEC (Executable file)
  Machine:                           Advanced Micro Devices X86-64
  Version:                           0x1
  Entry point address:               0x4003b0
  Start of program headers:          64 (bytes into file)
  Start of section headers:          7192 (bytes into file)
  Flags:                             0x0
  Size of this header:               64 (bytes)
  Size of program headers:           56 (bytes)
  Number of program headers:         9
  Size of section headers:           64 (bytes)
  Number of section headers:         34
  Section header string table index: 33
jo@ubuntu18:~/Desktop/linux64/gcc/10 memory$
```

图 8-3　readelf 标头

readelf 可以方便地浏览二进制可执行文件。图 8-4 展示了更多示例。

```
jo@ubuntu18:~/Desktop/linux64/gcc/10 memory$ readelf --symbols ./memory |grep main
     1: 0000000000000000     0 FUNC    GLOBAL DEFAULT  UND __libc_start_main@GLIBC_2.2.5 (2)
    64: 0000000000000000     0 FUNC    GLOBAL DEFAULT  UND __libc_start_main@@GLIBC_
    74: 00000000004004a0     0 NOTYPE  GLOBAL DEFAULT   11 main
jo@ubuntu18:~/Desktop/linux64/gcc/10 memory$
```

图 8-4　readelf 标记

我们使用 grep 表明要查找包含单词 main 的所有行。在这里，可以看到 main 函数从 0x4004a0 开始，就像我们在 GDB 中看到的那样。在以下示例中，我们在 symbols 表中查找标签 start 的每个匹配项。我们可以看到.data 段、.bss 段的起始地址，以及程序本身的起始地址，参见图 8-5。

```
jo@ubuntu18:~/Desktop/linux64/gcc/10 memory$ readelf --symbols ./memory |grep start
     1: 0000000000000000     0 FUNC    GLOBAL DEFAULT  UND __libc_start_main@GLIBC_2.2.5 (2)
     2: 0000000000000000     0 NOTYPE  WEAK   DEFAULT  UND __gmon_start__
    57: 00000000000600e50    0 NOTYPE  LOCAL  DEFAULT   16 __init_array_start
    61: 0000000000601018    0 NOTYPE  WEAK   DEFAULT   21 data_start
    64: 0000000000000000     0 FUNC    GLOBAL DEFAULT  UND __libc_start_main@@GLIBC_
    65: 0000000000601018    0 NOTYPE  GLOBAL DEFAULT   21 __data_start
    66: 0000000000000000     0 NOTYPE  WEAK   DEFAULT  UND __gmon_start__
    72: 00000000004003b0   43 FUNC    GLOBAL DEFAULT   11 _start
    73: 0000000000601051    0 NOTYPE  GLOBAL DEFAULT   22 __bss_start
jo@ubuntu18:~/Desktop/linux64/gcc/10 memory$
```

图 8-5 查看地址

下面使用指令查看一下内存情况，如下所示：

readelf --symbols ./memory |tail +10|sort -k 2 -r

tail 指令忽略了一些我们现在不感兴趣的行。我们按相反顺序对第二列(内存地址)进行排序。如你所见，掌握 Linux 命令的一些基本知识非常有用！

程序从某个低地址开始，而 main 的起始地址是 0x004004a0。查找.data 段 (0x00601018)的起始地址及其所有变量的地址，以及.bss(0x00601051)的起始地址和为其变量保留的地址。

在本章开始时，我们发现栈位于高内存中(请参阅 rsp)。使用 readelf，我们发现可执行代码位于内存的较低位置。在可执行代码之上，有.data 段和.bss 段。高位内存中的栈可能会增长；它朝着.bss 段向下增长。栈和其他段之间的可用空闲内存称为堆。

.bss 段中的内存是在运行时分配的；你可以很容易地进行查看。注意可执行文件的大小，然后更改以下内容：

    qvar   resq   3

更改为：

    qvar   resq   30000

重新编译程序，然后再次查看可执行文件的大小。大小将相同，因此在汇编/链接时不会保留额外的内存，参见图 8-6。

```
jo@ubuntu18:~/Desktop/linux64/gcc/10_memory$ readelf --symbols ./memory |tail +10|sort -k 2 -r
    70: 0000000000601098     0 NOTYPE  GLOBAL DEFAULT   22 _end
    71: 0000000000601071     1 OBJECT  LOCAL  DEFAULT   22 qvar
    49: 000000000060105d     1 OBJECT  LOCAL  DEFAULT   22 wvar
    49: 0000000000601059     1 OBJECT  LOCAL  DEFAULT   22 dvar
    48: 0000000000601058     1 OBJECT  LOCAL  DEFAULT   22 bvar
    75: 0000000000601058     0 OBJECT  GLOBAL HIDDEN    21 __TMC_END__
    35: 0000000000601054     1 OBJECT  LOCAL  DEFAULT   22 completed.7696
    22: 0000000000601054     0 SECTION LOCAL  DEFAULT   22
    73: 0000000000601051     0 NOTYPE  GLOBAL DEFAULT   22 __bss_start
    62: 0000000000601051     0 NOTYPE  GLOBAL DEFAULT   21 _edata
    47: 000000000060104d     1 OBJECT  LOCAL  DEFAULT   21 text2
    46: 0000000000601045     8 OBJECT  LOCAL  DEFAULT   21 qNum2
    45: 0000000000601041     1 OBJECT  LOCAL  DEFAULT   21 text1
    44: 0000000000601039     8 OBJECT  LOCAL  DEFAULT   21 qNum1
    43: 0000000000601035     4 OBJECT  LOCAL  DEFAULT   21 dNum
    42: 000000000060102b     2 OBJECT  LOCAL  DEFAULT   21 warray
    41: 0000000000601029     2 OBJECT  LOCAL  DEFAULT   21 wNum
    40: 0000000000601028     1 OBJECT  LOCAL  DEFAULT   21 bNum
    67: 0000000000601020     0 OBJECT  GLOBAL HIDDEN    21 __dso_handle
    21: 0000000000601018     0 SECTION LOCAL  DEFAULT   21
    61: 0000000000601018     0 NOTYPE  WEAK   DEFAULT   21 data_start
    65: 0000000000601018     0 NOTYPE  GLOBAL DEFAULT   21 __data_start
    20: 0000000000601000     0 SECTION LOCAL  DEFAULT   20
    59: 0000000000601000     0 OBJECT  LOCAL  DEFAULT   20 _GLOBAL_OFFSET_TABLE_
    19: 0000000000600ff0     0 SECTION LOCAL  DEFAULT   19
    18: 0000000000600e60     0 SECTION LOCAL  DEFAULT   18
    56: 0000000000600e60     0 OBJECT  LOCAL  DEFAULT   18 _DYNAMIC
    17: 0000000000600e58     0 SECTION LOCAL  DEFAULT   17
    36: 0000000000600e58     0 OBJECT  LOCAL  DEFAULT   17 __do_global_dtors_aux_fin
    55: 0000000000600e58     0 NOTYPE  LOCAL  DEFAULT   16 __init_array_end
    16: 0000000000600e50     0 SECTION LOCAL  DEFAULT   16
    38: 0000000000600e50     0 OBJECT  LOCAL  DEFAULT   16 __frame_dummy_init_array_
    57: 0000000000600e50     0 NOTYPE  LOCAL  DEFAULT   16 __init_array_start
    53: 0000000000400684     0 OBJECT  LOCAL  DEFAULT   15 __FRAME_END__
    15: 00000000004005d0     0 SECTION LOCAL  DEFAULT   15
    14: 00000000004005a4     0 SECTION LOCAL  DEFAULT   14
    58: 00000000004005a4     0 NOTYPE  LOCAL  DEFAULT   14 __GNU_EH_FRAME_HDR
    68: 00000000004005a0     4 OBJECT  GLOBAL DEFAULT   13 _IO_stdin_used
    13: 00000000004005a0     0 SECTION LOCAL  DEFAULT   13
    12: 0000000000400594     0 SECTION LOCAL  DEFAULT   12
    63: 0000000000400594     0 FUNC    GLOBAL DEFAULT   12 _fini
    60: 0000000000400590     2 FUNC    GLOBAL DEFAULT   11 __libc_csu_fini
    69: 0000000000400520   101 FUNC    GLOBAL DEFAULT   11 __libc_csu_init
    74: 00000000004004a0     0 NOTYPE  GLOBAL DEFAULT   11 main
    37: 0000000000400490     0 FUNC    LOCAL  DEFAULT   11 frame_dummy
    34: 0000000000400460     0 FUNC    LOCAL  DEFAULT   11 __do_global_dtors_aux
    33: 0000000000400420     0 FUNC    LOCAL  DEFAULT   11 register_tm_clones
    32: 00000000004003f0     0 FUNC    LOCAL  DEFAULT   11 deregister_tm_clones
    71: 00000000004003e0     2 FUNC    GLOBAL HIDDEN    11 _dl_relocate_static_pie
    72: 00000000004003b0    43 FUNC    GLOBAL DEFAULT   11 _start
```

图 8-6 readelf --symbols ./memory |tail +10|sort -k 2 -r 的输出

图 8-7 展示了加载可执行文件时内存的情况。

为什么了解内存结构很重要？ 重要的是要知道堆栈是向下生长的。当我们在本书的后面部分利用堆栈时，你将需要这些知识。另外，如果你要进行取证或恶意软件调查，那么分析内存也是一项必不可少的技能。这里仅涉及一些基本知识；如果你想了解更多信息，请参考前面提到的资源。

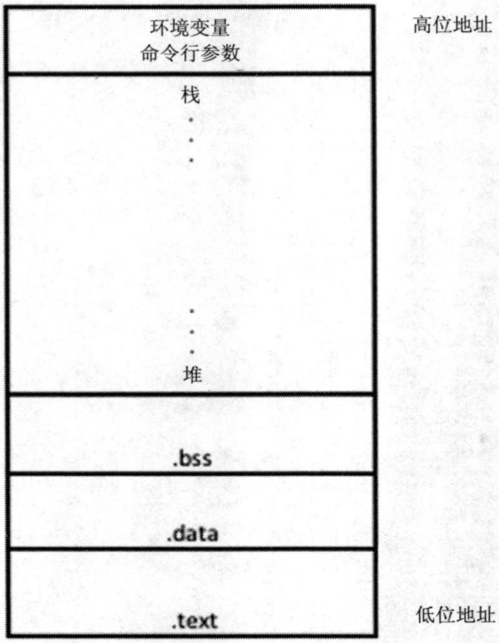

```
           环境变量              高位地址
           命令行参数
              栈
               .
               .
               .



               .
               .
               .
              堆

             .bss

             .data

             .text              低位地址
```

图 8-7　内存表

8.2　小结

本章内容：
- 处理内存的结构
- 如何避免无意中覆盖内存
- 如何使用 readelf 分析二进制代码

第 9 章

整数运算

本章将介绍许多关于整数的运算指令。浮点运算将在下一章中介绍。现在是时候快速复习一下第 2 章中有关二进制数的内容了。

9.1 从整数算术开始

代码清单 9-1 展示了我们将要分析的示例代码。

代码清单 9-1：icalc.asm

```
; icalc.asm
extern printf
section .data
    number1     dq 128      ; 用于执行数学计算的数字
    number2     dq 19
    neg_num     dq -12      ; 用于展示符号扩展
    fmt         db "The numbers are %ld and %ld",10,0
    fmtint      db "%s %ld",10,0
    sumi    db  "The sum is",0
    difi    db  "The difference is",0
    inci    db  "Number 1 Incremented:",0
    deci    db  "Number 1 Decremented:",0
    sali    db  "Number 1 Shift left 2 (x4):",0
    sari    db  "Number 1 Shift right 2 (/4):",0
    sariex  db  "Number 1 Shift right 2 (/4) with "
            db  "sign extension:",0
    multi   db  "The product is",0
```

```
        divi    db   "The integer quotient is",0
        remi    db   "The modulo is",0
section .bss
        resulti   resq 1
        modulo    resq 1
section .text
        global main
main:
        push  rbp
        mov   rbp,rsp
; 显示数字
        mov rdi, fmt
        mov rsi, [number1]
        mov rdx, [number2]
        mov rax, 0
        call printf
; 加法------------------------------------------------------
        mov rax, [number1]
        add rax, [number2]      ; 将 number2 添加到 rax
        mov [resulti], rax      ; 将总和移至 result(结果)
        ; 显示结果
        mov rdi, fmtint
        mov rsi, sumi
        mov rdx, [resulti]
        mov rax, 0
        call printf
; 减法------------------------------------------------------
        mov rax, [number1]
        sub  rax, [number2]     ; 从 rax 中减去 number2
        mov [resulti], rax
        ; 显示结果
        mov rdi, fmtint
        mov rsi, difi
        mov rdx, [resulti]
        mov rax, 0
        call printf
; 递增------------------------------------------------------
     mov rax, [number1]
     inc   rax        ; 将 rax 递增 1
```

```
        mov [resulti], rax
        ; 显示结果
              mov rdi, fmtint
              mov rsi, inci
              mov rdx, [resulti]
              mov rax, 0
              call printf
; 递减------------------------------------------------
    mov rax, [number1]
    dec   rax          ; 将 rax 递减 1
     mov [resulti], rax
    ; 显示结果
              mov rdi, fmtint
              mov rsi, deci
              mov rdx, [resulti]
              mov rax, 0
              call printf
; 左移运算----------------------------------------------
    mov rax, [number1]
    sal   rax, 2       ; 将 rax 乘以 4
    mov [resulti], rax
    ; 显示结果
              mov rdi, fmtint
              mov rsi, sali
              mov rdx, [resulti]
              mov rax, 0
              call printf
; 右移运算----------------------------------------------
    mov rax, [number1]
    sar   rax, 2       ; 将 rax 除以 4
    mov [resulti], rax
    ; 显示结果
              mov rdi, fmtint
              mov rsi, sari
              mov rdx, [resulti]
              mov rax, 0
              call printf
; 带符号扩展的右移运算------------------------------------
    mov rax, [neg_num]
```

```asm
        sar  rax, 2         ; 将 rax 除以 4
    mov [resulti], rax
    ; 显示结果
            mov rdi, fmtint
            mov rsi, sariex
            mov rdx, [resulti]
            mov rax, 0
            call printf
; 乘法----------------------------------------------------------
    mov rax, [number1]
    imul qword [number2]    ; 将 rax 与 number2 相乘
    mov [resulti], rax
    ; 显示结果
            mov rdi, fmtint
            mov rsi, multi
            mov rdx, [resulti]
            mov rax, 0
            call printf
; 除法----------------------------------------------------------
    mov rax, [number1]
    mov rdx, 0              ; rdx 需要在执行 idiv 指令之前为 0
    idiv qword [number2]    ; 用 rax 除以 number2，模存放在 rdx
    mov [resulti], rax
    mov [modulo], rdx       ; 将 rdx 中的数据(模)移到 modulo
    ; 显示结果
        mov rdi, fmtint
            mov rsi, divi
            mov rdx, [resulti]
            mov rax, 0
            call printf
            mov rdi, fmtint
            mov rsi, remi
            mov rdx, [modulo]
            mov rax, 0
            call printf
    mov rsp,rbp
    pop rbp
    ret
```

输出如图 9-1 所示。

```
jo@UbuntuDesktop:~/Desktop/linux64/gcc/11 icalc$ make
nasm -f elf64 -g -F dwarf icalc.asm -l icalc.lst
gcc -o icalc icalc.o -no-pie
jo@UbuntuDesktop:~/Desktop/linux64/gcc/11 icalc$ ./icalc
The numbers are 128 and 19
The sum is 147
The difference is 109
Number 1 Incremented: 129
Number 1 Decremented: 127
Number 1 Shift left 2 (x4): 512
Number 1 Shift right 2 (/4): 32
Number 1 Shift right 2 (/4) with sign extension: -3
The product is 2432
The integer quotient is 6
The modulo is 14
jo@UbuntuDesktop:~/Desktop/linux64/gcc/11 icalc$
```

图 9-1 整数运算

9.2 分析算术指令

算术指令有很多，此处将展示其中的一部分，其他部分与你将在此处学到的类似。在研究算术指令前，请注意，我们将 printf 与两个以上的参数一起使用，因此需要一个额外的寄存器：第一个参数放入 rdi，第二个参数放入 rsi，第三个参数放入 rdx。这就是 printf 期望我们在 Linux 中提供参数的方式。稍后谈论调用约定时，你将了解更多。

以下是一些算术指令：

- 第一条指令是 add，可用于有符号或无符号整数的加法运算。将第二个操作数(源)与第一个操作数(目标)相加，并将结果放在第一个操作数(目标)中。目标操作数可以是寄存器或内存位置。源可以是即时值、寄存器或内存位置。源和目标不能是同一指令中的内存位置。当结果总和太大而无法放入目标时，将为有符号整数设置 CF 标志。对于无符号整数，设置 OF 标志。结果为 0 时，ZF 标志设置为 1；结果为负值时，设置 SF 标志。
- 减法指令 sub 类似于加法指令。
- 要将一个寄存器或内存位置的值加 1，请使用 inc 指令。类似地，dec 可用于将内存位置的值或寄存器减 1。
- 移位运算是一种特殊指令。sal 表示左移，实际上是成倍增加；如果向左移动一个位置，则乘以 2。每向左移动一位，最右边添加一个 0。取二进制数字 1。向左移动一位，你将获得以二进制表示的 10 或十进制表示的 2。再次向左移一位，则二进制表示形式为 100 或二进制表示的 4。如果向左移动两个位置，则乘以 4。如果要乘以 6，该怎么办？你可以按此顺序向左移动两次，然后按顺序将原始源添加两次。

- sar 表示右移,与左移相似,但是它意味着除以 2。每一个位都向右移动一位,并在左边增加一个附加位(0)。然而,这里有一个复杂的问题:如果原始值是负数,则最左边的位将是 1;如果移位指令在左边加上一个 0,则该值将变为正数,结果将是错误的。因此,在负值的情况下,sar 将在左侧添加一个 1,在正值的情况下,将在左侧添加一个 0。这称为符号扩展。顺便说一句,查看十六进制数是否为负的一种快速方法是查看字节 7(最左边的字节,从最右边的字节 0 开始计数)。如果字节 7 以 8、9、A、B、C、D、E 或 F 开头,则数字为负数。但你需要考虑所有 8 个字节。例如,0xd12 仍然是正数,因为最左边的字节(未显示)是 0。
- 也有非算术位移指令,我们将在第 16 章讨论它们。
- 接下来,我们将整数相乘。对于无符号整数的乘法,可以使用 mul 指令,对于有符号的乘法需要使用 imul。本例中我们将使用 imul,即有符号乘法,它提供了更大的灵活性;imul 可以使用一个、两个或三个操作数。在我们的示例中,使用一个操作数;imul 指令后面的操作数与 rax 中的值相乘。你可能希望乘积的结果存储在 rax 中,但这并不完全正确。下面用一个例子来说明:当你用一个三位数的数字乘两位数时,乘积为四位或五位数。将 48 位数字与 30 位数字相乘时,将获得 77 位数字或 78 位数字,并且该值不适合 64 位寄存器。为解决这个问题,指令 imul 将结果乘积的低 64 位存储在 rax 中,并将高 64 位存储在 rdx 中。这可能非常具有欺骗性!下面做一个实验:回到 SASM 中的源代码。修改 number1 使其值改为 12345678901234567,并修改 number2 使其值改变为 100。这两个数的乘积仅适合 rax,你可以在 SASM 调试模式下进行检查。在 imul 指令之前设置一个断点。重新启动调试模式,并逐步执行该程序。乘法的结果将是 1234567890123456700,正如执行 imul 指令后在 rax 中看到的那样。现在将 number2 修改为 10000。重新启动调试。执行 imul 后再看 rax。你会看到乘积是很大的负数! 这是因为 rax 中的最高有效位是 1,并且 SASM 确定该位必须为负数。另外,printf 认为 rax 是负数,因为 rax 最左边的位是 1,因此它被假定为负数。所以,请谨慎使用 printf!

我们将进行更深入的研究:一旦执行了 imul 指令,rax 的值就变为 0xb14e9f812f364970。二进制形式是 1011000101001110100111111100000010010111100110110010010010010,最高有效位为 1,因此为负数。

rdx 中的值将是 0x6。那就是 000000000000000000000000000000000000000000000000000000000000110,最高有效位为 0,因此是正数。

实际乘积为 0x6b14e9f812f364970，可按以下顺序组合 rdx 和 rax 来获得正确的乘积：rdx:rax。如果将此十六进制数转换为十进制数，则乘积为 123456789012345670000。请参见图 9-2。

图 9-2　rax 和 rdx 中的值

可在互联网上找到十六进制到十进制的转换应用程序；参见 https://www.rapidtables.com/convert/number/hexto-decimal.html。

- 下面继续进行整数除法 idiv。实际上，这与乘法相反。将 rdx:rax 中的被除数除以源操作数中的除数，然后将整数结果存储在 rax 中。模(余数)可以在 rdx 中找到。请务必在每次使用 idiv 之前将 rdx 设置为零，否则所得商数可能是错误的。

64 位整数乘法和除法有些微妙之处，你可在英特尔手册中找到更多详细信息。这里仅给出了概述，可作为对整数运算的一般性介绍。在英特尔手册中，你不仅能找到有关指令的详细信息，还能找到可在特定情况下使用的其他大量算术指令。

9.3　小结

本章内容：
- 如何执行整数运算
- 如何执行左/右移位运算

- 使用 rax 和 rdx 存储乘积
- 使用 rax 和 rdx 执行除法运算
- 当使用 printf 打印值时要小心

第 10 章

堆栈

我们已经讨论过寄存器，一种快速临时存储类型，寄存器可用于在指令执行期间存储使用的值或地址。另一种是速度较慢的存储——内存，处理器可以在内存中将值存储更长时间。然后是堆栈(Stack)，一种连续的内存位置数组。

10.1 理解堆栈

如第 8 章所述，堆栈从高位内存开始，并且当它增长时，是朝下增长，就像挂着的冰柱增长时一样，向下增长。使用 push 指令将对象放置在堆栈上，使用弹出指令将对象从堆栈中删除。每次执行 push，堆栈就会增加。每次执行 pop，堆栈就会减小。你可通过监视 rsp(栈指针)来验证堆栈的这种行为，它指向堆栈的顶部(实际上是底部，因为它向下增长)。

堆栈可以用作临时存储，以将值保存在寄存器中并在以后调用它们，或更重要的是，可以将值传递给函数。函数或程序将在后面详细介绍。

在代码清单 10-1 的示例代码中，我们将使用堆栈来反转字符串。

代码清单 10-1：stack.asm

```
; stack.asm
extern printf
section .data
    strng       db      "ABCDE",0
    strngLen    equ     $ - strng-1     ; 字符串的长度(不包括终止 0)
    fmt1        db      "The original string: %s",10,0
    fmt2        db      "The reversed string: %s",10,0
section .bss
```

```asm
section .text
    global main
main:
push rbp
mov    rbp,rsp

; 打印原始字符串
    mov   rdi, fmt1
    mov   rsi, strng
    mov   rax, 0
    call  printf

;将字符串逐个字符地压入堆栈
    xor   rax, rax
    mov   rbx, strng        ; 将字符串的地址放入 rbx
    mov   rcx, strngLen     ; 将长度放入 rcx
    mov   r12, 0            ; 使用 r12 作为指针
    pushLoop:
        mov   al, byte [rbx+r12]  ; 把字符移到 rax
        push rax                   ; 把 rax 压入堆栈
        inc   r12                  ; 将字符指针增加 1
        loop pushLoop              ; 继续循环

;将字符串从堆栈中逐字符弹出
;这将反转原始字符串
    mov rbx, strng          ; strng 的地址放入 rbx
    mov rcx, strngLen       ; 长度放入 rcx
    mov r12, 0              ; 使用 r12 作为指针
    popLoop:
        pop rax                    ; 从堆栈上弹出一个字符
        mov byte [rbx+r12], al    ; 将字符移到 strng
        inc   r12                  ; 将指针增加 1
        loop popLoop               ; 继续循环
        mov byte [rbx+r12],0      ; 使用 0 终止字符串

; 打印反转后的字符串
    mov   rdi, fmt2
    mov   rsi, strng
    mov   rax, 0
    call  printf
```

```
mov    rsp,rbp
pop    rbp
ret
```

图 10-1 展示了输出结果。

```
jo@UbuntuDesktop:~/Desktop/linux64/gcc/12 stack$ make
nasm -f elf64 -g -F dwarf stack.asm -l stack.lst
gcc -o stack stack.o
jo@UbuntuDesktop:~/Desktop/linux64/gcc/12 stack$ ./stack
The original string: ABCDE
The reversed string: EDCBA
jo@UbuntuDesktop:~/Desktop/linux64/gcc/12 stack$
```

图 10-1　反转一个字符串

首先，请注意，为了计算字符串长度，我们将字符串的长度减少 1，忽略了终止的 0。否则，反转的字符串将以 0 开头。然后显示原始字符串，之后是一个新行。我们将使用 rax 来压入字符，所以下面首先使用 xor 初始化 rax。字符串的地址放入 rbx，我们将使用循环指令，因此将 rcx 设置为字符串长度。然后使用循环从第一个字符开始将字符逐个压入堆栈。我们将一个字符(字节)移到 al 中。然后将 rax 压入堆栈。每次使用 push 时，会将 8 个字节压入堆栈中。如果之前没有初始化过 rax，很可能 rax 包含高字节的值，而将这些值压入堆栈可能不是我们想要的。之后，堆栈包含压入的字符以及 al 上的附加 0 位。

循环结束后，最后一个字符位于堆栈"顶部"，实际上，该字符位于"冰柱"的最低地址，因为堆栈向下增长。开始另一个循环，该循环从堆栈中一个接一个地弹出字符，并将它们一个接一个地存储在内存中的原始字符串中。请注意，我们只需要 1 个字节，因此我们弹出到 rax 时使用 al。

下面是正在发生的事情的概述(参见图 10-2)。右边是原始字符串，字符被一个接一个地压入堆栈(被附加到堆栈上)。之后，字符被弹出并发送回字符串的内存地址，由于堆栈的"后进先出"工作机制，所以字符串被反转。

你必须以某种方式跟踪堆栈上的压入内容和顺序。例如，当你使用堆栈临时存储寄存器时，请确保以相反的正确顺序弹出寄存器；否则，你的程序将出现错误，或者在最坏的情况下可能会崩溃。也就是说，当你按以下顺序压入时：

```
push rax
push rbx
push rcx
```

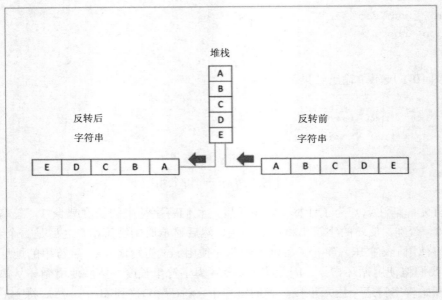

图 10-2　反转一个字符串的模式

然后，根据"后进先出"原则，你必须按如下方式弹出：

```
pop rcx
pop rbx
pop rax
```

除了寄存器之外，你还可以压入内存和即时值。可以弹出到寄存器或内存位置，但不能弹出到一个即时值，这非常明显。

可以使用 pushf 和 popf 指令将标志寄存器压入或弹出堆栈。

10.2　跟踪堆栈

跟踪堆栈很重要，而我们的老朋友 DDD 具有一些简单的功能可以做到这一点。首先，在编辑器中打开源代码，并删除 SASM 添加的调试行。然后保存文件并退出。在命令行上编译程序，然后输入以下内容：

```
ddd stack
```

在菜单栏中选择 Data | Status Displays，然后向下滚动，直至找到 Backtrace of the stack 并启用它。例如，在 main:处设置一个断点，然后在浮动面板中单击 Run。现在开始调试，并使用 step 按钮单步执行程序(相信你不希望通过 printf 函数逐行执

行)。在上方的窗口中查看堆栈的显示和更新方式。不必担心显示的初始内容。当你到达 push 指令之后的指令时，会看到字符以 ASCII 十进制表示(41、42 等)被压入堆栈。观察在第二个循环中堆栈如何减少。这是查看堆栈上的内容和顺序的简单方法。

图 10-3 展示了大致外观。

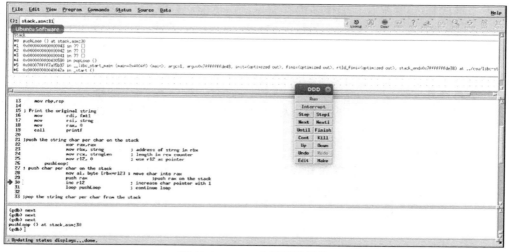

图 10-3　DDD 中显示的堆栈

如前所述，DDD 是开源的且已过时。无法保证它在将来继续按预期工作，但目前而言，它虽然不是很优雅，但基本可以满足使用需求。

你也可以强制 SASM 显示堆栈，但这需要更多的手工操作。它的工作原理是这样的：你可在 SASM 中进行调试时显示内存变量，堆栈只是内存位置的列表，而 rsp 指向最低位置。因此，必须让 SASM 展示地址 rsp 和内存位置之上的内容。图 10-4 展示了 SASM 中显示堆栈的示例内存窗口。

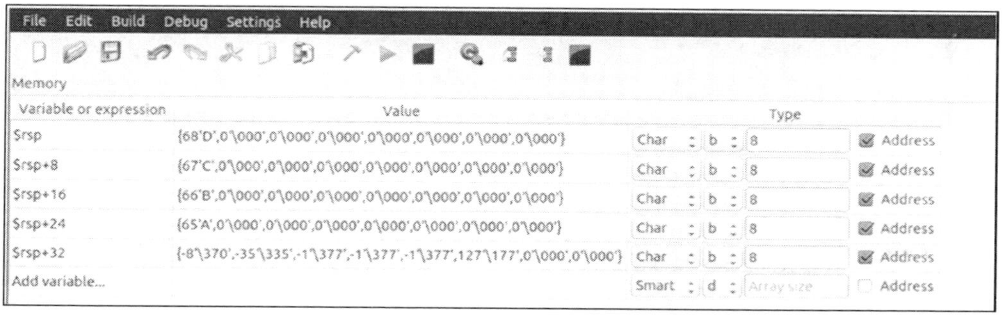

图 10-4　SASM 中显示的堆栈

我们将 rsp 称为$rsp。每次会将堆栈地址增加 8($rsp + 8)，因为每次 push 都将 8

个字节压入堆栈。对于类型，我们指定字符、字节、8 个字节和地址。选择字符是因为我们要压入一个字符串，这样易于阅读；选择字节是因为我们对字节值感兴趣(al 每次包含 1 个字节)，因此每次都压入 8 个字节。rsp 包含一个地址。单步执行该程序，然后查看堆栈如何改变。

它可以正常工作，但你必须手动详细说明每个堆栈的内存位置，如果你使用的是大型堆栈和/或要跟踪其他很多内存变量，这可能十分麻烦。

10.3　小结

本章内容：
- 堆栈从高位内存中的一个地址开始，然后增长到较低的地址。
- 递减栈指针(rsp)。
- 弹出栈指针(rsp)。
- 以相反的顺序进行压入和弹出。
- 如何使用 DDD 查看堆栈。
- 如何使用 SASM 查看堆栈。

第 11 章

浮点运算

你已经了解整数算术，现在介绍一些浮点运算。没有什么难理解的内容，浮点值中有一个小数点和零个或多个小数。有两种浮点数：单精度和双精度。双精度更精确，因为它可以处理更多有效数字。有了这些信息，你现在已经足够了解如何运行和分析本章中的示例程序了。

11.1 单精度与双精度

对于那些更好奇的人来说，故事是这样的。

单精度数字存储在 32 位中：1 个符号位，8 个指数位和 23 个小数位。

```
S    EEEEEEEE   FFFFFFFFFFFFFFFFFFFFFFF
0    1      8   9                     31
```

双精度数字存储在 64 位中：1 个符号位，11 个指数位和 52 个小数位。

```
S    EEEEEEEEEEE   FFFFFFFFFFFF......FFFFFFFFF
0    1         11  12                        63
```

符号位很简单。当数字为正数时，符号位是 0。当数字为负数时，符号位为 1。指数位更复杂。下面看一个十进制的例子。

```
200 = 2.0 × 10²
5000.30 = 5.0003 × 10³
```

这是一个二进制示例：

```
1101010.01011 = 1.0101001011 x 2⁶  (将小数点向左移动六个位置)
```

指数可以是正、负或零。为了明确这一区别，在单精度情况下，在存储正指数之前将 127 加到正指数上。这意味着零指数将存储为 127。127 被称为偏差。使用双精度值时，偏差为 1023。

在上例中，1.0101001011 被称为有效位或尾数。根据假设，有效位的第一位是 1(是标准化的)，因此不会存储。

下面是一个简单例子，用来说明它是如何工作的。可使用 https://babbage.cs.qc.cuny.edu/IEEE-754/进行验证和实验。

单精度，十进制数 10：

- 十进制数 10 的二进制形式是 1010。
- 符号位为 0，因为数字是正数。
- 获取格式为 b.bbbb 的数字。1.010 是有效位数，根据需要以 1 开头。前导 1 将不被存储。
- 因此，指数是 3，因为我们将小数点移动了三个位置。因为指数为正，所以加了 127，得到 130，二进制值为 10000010。
- 因此，十进制单精度数字 10 将存储为：

```
0  10000010   01000000000000000000000
S  EEEEEEEE   FFFFFFFFFFFFFFFFFFFFFFF
```

或 41200000(十六进制)。

注意，相同值的十六进制表示形式在单精度和双精度方面有所不同。为什么不总是使用双精度并从高精度中受益呢？ 双精度计算比单精度计算要慢，并且操作数占用更多内存。

如果你认为这很复杂，你是对的。在互联网上找一个合适的工具来完成或至少对转换进行验证。

在较旧的程序中，你可能遇到 80 位浮点数，这些数字具有自己的指令，称为 FPU 指令。这是过去的遗留功能，不应在新的开发中使用。但是你会时不时地在互联网上的文章中看到 FPU 指令。

下面来做一些有趣的事情。

11.2　浮点数编程

代码清单 11-1 展示了示例程序。

代码清单 11-1：fcalc.asm

```
; fcalc.asm
```

```nasm
        extern printf
        section .data
            number1     dq  9.0
            number2     dq  73.0
            fmt         db  "The numbers are %f and %f",10,0

            fmtfloat    db  "%s %f",10,0
            f_sum       db  "The float sum of %f and %f is %f",10,0
            f_dif       db  "The float difference of %f and %f is %f",10,0
            f_mul       db  "The float product of %f and %f is %f",10,0
            f_div       db  "The float division of %f by %f is %f",10,0
            f_sqrt      db  "The float squareroot of %f is %f",10,0
        section .bss
        section .text
            global main
main:
        push    rbp
        mov     rbp,rsp
; 打印数字
        movsd   xmm0, [number1]
        movsd   xmm1, [number2]
        mov     rdi,fmt
        mov     rax,2       ; 两个浮点数
        call    printf
; 和(sum)
        movsd   xmm2, [number1]     ; 双精度浮点数放入 xmm
        addsd   xmm2, [number2]     ; 将一个双精度浮点数加到 xmm 中
            ; 打印结果
        movsd   xmm0, [number1]
        movsd   xmm1, [number2]
        mov     rdi,f_sum
        mov     rax,3       ; 三个浮点数
        call    printf
; 差(difference)
        movsd   xmm2, [number1]     ; 双精度浮点数放入 xmm
        subsd   xmm2, [number2]     ; 从 xmm 中减去一个双精度浮点数
            ; 打印结果
        movsd   xmm0, [number1]
        movsd   xmm1, [number2]
```

```
            mov     rdi,f_dif
            mov     rax,3       ;三个浮点数
            call    printf
; 积(multiplication)
    movsd   xmm2, [number1]     ; 双精度浮点数放入 xmm
    mulsd   xmm2, [number2]     ; 将 xmm 中的双精度浮点数乘以一个双精度浮点数
            ; 打印结果
            mov     rdi,f_mul
            movsd   xmm0, [number1]
            movsd   xmm1, [number2]
            mov     rax,3       ;三个浮点数
            call    printf
; 除法(division)
    movsd   xmm2, [number1]     ; 双精度浮点数放入 xmm
    divsd   xmm2, [number2]     ; 除以 xmm0
            ; 打印结果
            mov     rdi,f_div
            movsd   xmm0, [number1]
            movsd   xmm1, [number2]
            mov     rax,1       ;一个浮点数
            call    printf
; 平方根(squareroot)
    sqrtsd  xmm1, [number1]     ; 双精度浮点数放入 xmm
            ; 打印结果
            mov     rdi,f_sqrt
            movsd   xmm0, [number1]
            mov     rax,2       ;两个浮点数
            call    printf
; exit
            mov rsp, rbp
            pop rbp             ;撤销一开始的压入
            ret
```

这是一个简单的程序；实际上，打印要比浮点计算耗费更多精力。

使用调试器单步执行程序并查看寄存器和内存；例如，查看 9.0 和 73.0 如何存储在内存地址 number1 和 number2 中。这些是双精度浮点值。

请记住，在 SASM 中进行调试时，xmm 寄存器位于寄存器窗口的底部，即 ymm 寄存器的最左侧。

movsd 的意思是"移动一个双精度浮点值"。使用 movss 移动单精度值。类似

地，还有 addss、subss、mulss、divss 和 sqrtss 指令。

剩下的都非常简单了！输出如图 11-1 所示。

```
jo@UbuntuDesktop:~/Desktop/linux64/gcc/13 fcalc$ make
nasm -f elf64 -g -F dwarf fcalc.asm -l fcalc.lst
gcc -o fcalc fcalc.o
jo@UbuntuDesktop:~/Desktop/linux64/gcc/13 fcalc$ ./fcalc
The numbers are 9.000000 and 73.000000
The float sum of 9.000000 and 73.000000 is 82.000000
The float difference of 9.000000 and 73.000000 is -64.000000
The float product of 9.000000 and 73.000000 is 657.000000
The float division of 9.000000 by 73.000000 is 0.123288
The float squareroot of 9.000000 is 3.000000
jo@UbuntuDesktop:~/Desktop/linux64/gcc/13 fcalc$
```

图 11-1　fcalc.asm 的输出

现在你已经了解了堆栈，请尝试以下操作：在开始处注释掉 push rbp，在结尾处弹出 rbp。编译并运行程序，看看会发生什么：程序会崩溃！崩溃的原因将在后面进行说明，这与堆栈对齐有关。

11.3　小结

本章内容：
- 使用 xmm 寄存器进行简单的浮点计算
- 单精度浮点数与双精度浮点数的差别
- 指令 movsd、addsd、subsd、mulsd、divsd 和 sqrtsd

第 12 章

函数

汇编语言不是一种"结构化语言"。看看大量的 jmp 指令和标签，它们使程序执行可以来回跳转。现代高级编程语言具有 do ... while、while ... do、case 等结构。然而汇编语言不是这样的。

但与现代程序语言类似，汇编语言通过函数和过程给代码赋予更多结构。函数执行指令并返回一个值；过程执行指令但不返回值。

在本书中，我们已经使用过函数，即名为 printf 的外部函数，它是一个 C 库函数。在本章中，我们将介绍简单的函数；在后续章节中，将介绍函数的重要方面，例如堆栈对齐、外部函数和调用约定。

12.1 编写一个简单的函数

代码清单 12-1 展示了一个汇编程序的示例，其中有一个简单的函数用来计算圆的面积。

代码清单 12-1：function.asm

```
; function.asm
extern printf
section .data
    radius    dq    10.0
    pi        dq    3.14
    fmt       db    "The area of the circle is %.2f",10,0
section .bss
section .text
    global main
```

```
;--------------------------------------------
main:
push rbp
mov   rbp, rsp
    call    area              ; 调用函数
    mov     rdi,fmt           ; 打印格式
    movsd   xmm1, [radius]    ; 将浮点数放入 xmm1
    mov     rax,1             ; area(面积)在 xmm0 中
    call    printf
leave
ret
;--------------------------------------------
area:
push rbp
mov   rbp, rsp
    movsd   xmm0, [radius]    ; 将浮点数放入 xmm0
    mulsd   xmm0, [radius]    ; 用浮点数乘以 xmm0
    mulsd   xmm0, [pi]        ; 用浮点数乘以 xmm0
leave
ret
```

图 12-1 显示了输出。

```
jo@UbuntuDesktop:~/Desktop/linux64/gcc/14 function$ make
nasm -f elf64 -g -F dwarf function.asm -l function.lst
gcc -o function function.o
jo@UbuntuDesktop:~/Desktop/linux64/gcc/14 function$ ./function
The area of the circle is 314.00
jo@UbuntuDesktop:~/Desktop/linux64/gcc/14 function$
```

图 12-1　function.asm 的输出

　　main 部分和以前一样用标签 main 标识，然后有一个函数，用标签 area 标识。在 main 中，函数 area 被调用，它使用 radius 和 pi 来计算圆的面积，这是存储在内存中某个位置的变量。如你所见，函数必须有一个序言和一个尾声，类似于 main。

　　计算出的面积存储在 xmm0 中。从 area 函数返回 main 时，printf 被调用，rax 包含值 1，这意味着需要打印一个 xmm 寄存器。这里介绍一个新指令：leave。该指令的作用与 mov rsp, rbp 和 pop rbp(尾声)相同。

　　如果函数有一个返回值，则将 xmm0 用作浮点值，将 rax 用作其他值，例如整数或地址。函数参数 pi 和 radius 位于内存中。但最好使用寄存器和堆栈来存储函数参数。使用内存变量将值传递给函数会在 main 函数和其他函数中使用的值之间产生命名冲突，并使代码的"可移植性"降低。

12.2 更多函数

下面使用另一个示例来讨论函数的更多特征(参见代码清单 12-2)。

代码清单 12-2：function2.asm

```
; function2.asm
extern printf
section .data
        radius    dq    10.0
section .bss
section .text
;--------------------------------------------
area:
        section .data
                .pi   dq    3.141592654    ; area 的局部变量
        section .text
push rbp
mov rbp, rsp
    movsd  xmm0, [radius]
    mulsd  xmm0, [radius]
    mulsd  xmm0, [.pi]
leave
ret
;--------------------------------------------
circum:
section .data
    .pi  dq 3.14         ; circum 的局部变量
section .text
push   rbp
mov    rbp, rsp
    movsd  xmm0, [radius]
    addsd  xmm0, [radius]
    mulsd  xmm0, [.pi]
leave
ret
;--------------------------------------------
circle:
section .data
```

```
        .fmt_area    db "The area is %f",10,0
        .fmt_circum  db "The circumference is %f",10,0
section .text
push rbp
mov  rbp, rsp
    call area
    mov  rdi,.fmt_area
    mov  rax,1      ; area(面积)存放在 xmm0 中
    call printf
    call circum
    mov  rdi,.fmt_circum
    mov  rax,1      ; circumference(周长)存放在 xmm0 中
    call printf
leave
ret
;---------------------------------------------
    global main
main:
push rbp
mov  rbp, rsp
    call circle
leave
ret
```

这里，主要调用函数 circle，该函数依次调用函数 area 和 circum。所以，函数可以调用其他函数。实际上，main 只是一个调用其他函数的函数。但是要注意函数不能嵌套，这意味着函数不能包含其他函数的代码。

此外，函数可以有自己的段，例如.data、.bss 和.text。pi 和 fmt 变量之前的句点表示局部变量，这意味着该变量只适用于声明它的函数。在函数区域中，我们使用了一个与函数 circum 中使用的 pi 不同的 pi 值。在 main 的.data 段中声明的变量 radius 在这个源代码列表中的每个函数中都是可见的，包括 main。建议尽可能使用局部变量；这样可以降低变量名冲突的风险。

图 12-2 展示了该程序的输出。

```
jo@UbuntuDesktop:~/Desktop/linux64/gcc/15 function2$ make
nasm -f elf64 -g -F dwarf function2.asm -l function2.lst
gcc -g -o function2 function2.o
jo@UbuntuDesktop:~/Desktop/linux64/gcc/15 function2$ ./function2
The area is 314.16
The circumference is 62.80
jo@UbuntuDesktop:~/Desktop/linux64/gcc/15 function2$
```

图 12-2 function2.asm 的输出

12.3 小结

本章内容：
- 如何使用函数
- 函数可以拥有自己的 .data 和 .bss 段
- 函数不能嵌套
- 函数可以调用其他函数
- main 只是一个函数而已
- 如何使用本地变量

第 13 章

栈对齐和栈帧

当主程序调用一个函数时,它会在栈上压入一个 8 字节的返回地址。该 8 字节地址是函数之后要执行的指令的地址。因此,当函数结束时,程序执行将从栈中找到返回地址,并在函数调用后继续操作。在函数内部,我们还可将栈用于不同目的。每次将某些内容压入栈时,栈指针将减少 8 个字节;每次从栈中弹出内容时,栈指针将增加 8 个字节。因此,我们必须确保在离开函数之前将栈"恢复"到适当的值。否则,正在执行的程序将在函数调用后为要执行的指令提供错误地址。

13.1 栈对齐

在英特尔手册中,你会发现当调用函数时,栈要求必须是 16 字节对齐。这听起来有点奇怪,因为栈是在 8 字节(或 64 位)内存中构建的。原因是有一些 SIMD 指令对较大的数据块执行并行操作,并且这些 SIMD 指令可能要求这些数据位于内存中 16 字节倍数的地址上。在前面的示例中,将 printf 与 xmm 寄存器一起使用时,会将栈对齐为 16 个字节,而没有明确告诉你。回到第 11 章的浮点运算,我们通过注释掉 push rbp 和 pop rbp 使程序崩溃。程序崩溃是因为删除这些指令导致栈未对齐。如果你使用不带 xmm 寄存器的 printf,则不必进行栈对齐即可解决问题,但是如果这样做,总有一天会出问题。

后续章节将讨论 SIMD 和对齐方式,因此,如果前面的说明对你没有意义,也不必担心。现在,请记住,在调用函数时,需要将栈对齐为 16 字节倍数的地址。

就处理器而言,main 只是另一个函数。在程序开始执行之前,栈是对齐的。在 main 启动之前,一个 8 字节的返回地址被压入栈,这意味着栈在 main 启动时没有对齐。如果在 main 的开始与函数的调用之间未使用栈,则栈指针 rsp 不会对齐 16 字节。你可以通过查看 rsp 来验证:如果 rsp 以 0 结尾,则它是 16 位对齐的。要使其为零，可将某些内容压入栈，以使其变成 16 位对齐。当然，不要忘记相应的弹出指令。

这种对齐要求是使用序言和尾声的原因之一。main 和函数中的第一条指令应将某些内容压入栈以使其对齐。这就是使用序言指令 push rbp 的原因。rbp 寄存器也称为基指针。

我们为什么要使用 rbp？在序言中，当使用栈帧(稍后说明)时修改了 rbp，因此在将 rbp 用于栈帧之前，将其压入栈以在返回时将其保留。即使不构建栈帧，rbp 也是对齐栈的理想选择，因为它不用于向函数传递参数。参数传递将在本章后面讨论。在序言中，我们还使用了 mov rbp, rsp 指令。该指令保留 rsp，rsp 是包含返回地址的栈指针。序言和尾声是相反的。最好不要干预 rbp！在后续章节中，你将看到其他许多用来对齐栈的方法。

代码清单 13-1 展示了一些可以使用的源代码。使用 SASM 调试和逐步执行程序时，请注意 rsp。注释掉 push rbp 和 pop rbp，看看会发生什么。如果栈没有对齐，当程序执行到 printf 时将崩溃。那是因为 printf 肯定需要栈对齐。

在此程序中，我们不使用完整的序言和结尾。也就是说，我们不构建栈帧。我们仅使用 push 和 pop 来说明对齐方式。

代码清单 13-1：aligned.asm

```
; aligned.asm
extern printf
section .data
     fmt   db "2 times pi equals %.14f",10,0
     pi  dq 3.14159265358979
section .bss
section .text
;----------------------------------------------
func3:
     push rbp
          movsd  xmm0, [pi]
          addsd  xmm0, [pi]
          mov   rdi,fmt
          mov   rax,1
          call   printf       ; 打印一个浮点数
     pop rbp
     ret
;----------------------------------------------
func2:
```

```
        push rbp
                call func3       ; 调用第三个函数
        pop   rbp
        ret
;-------------------------------------------
func1:
        push rbp
                call  func2      ; 调用第二个函数
        pop   rbp
        ret
;-------------------------------------------
        global main
main:
        push rbp
                call func1       ; 调用第一个函数
        pop   rbp
ret
```

请注意，如果执行一定数量的调用(偶数或奇数，取决于启动方式)，即使压入/弹出不对齐，栈也会对齐16字节，并且程序不会崩溃。纯属运气！

13.2 有关栈帧的更多信息

可以区分两种类型的函数：分支(branch)函数和叶(leaf)函数。分支函数包含对其他函数的调用，而叶函数仅执行一些命令，然后返回到父函数而不必调用任何其他函数。

原则上，每次调用函数时，都需要构建一个栈帧。具体操作如下：在被调用的函数中，首先在16字节的边界上对齐栈，即 push rbp。然后将栈指针 rsp 保存到 rbp 中。退出函数时，还原 rsp 并弹出 rbp 以还原 rbp。这就是函数序言和尾声的作用。在函数内部，寄存器 rbp 现在充当原始栈位置的锚点。每当一个函数调用另一个函数时，新函数应构建自己的栈帧。

在叶函数中，通常可以忽略栈帧和栈对齐；只要你不把栈搞乱就可以。请注意，当你在函数中调用 printf 等时，该函数不是叶函数。同样，如果你的函数不使用 SIMD 指令，则不必担心对齐问题。

编译器具有优化功能，有时当你查看由编译器生成的代码时，会发现没有使用栈帧。当编译器在优化过程中注意到不需要栈帧时，就会发生这种情况。

无论如何，始终包括一个栈帧并检查堆栈对齐是一个好习惯。以后可为你省去很多麻烦。包含栈帧的一个很好的理由是，GDB 和基于 GDB 的调试器(例如 DDD 和 SASM)都期望可以找到一个栈帧。如果代码中没有栈帧，调试器将发生不可预测的行为，例如忽略断点或跳过指令。从上一章中获取一些代码(如 alife.asm)，注释掉函数序言和尾声，然后启动 GDB 并查看会发生什么。

作为附加练习，使用 SASM 或 GDB 查看上一章中的代码(function2.asm)，并查看栈在执行期间如何保持对齐。

这里有一个附加的快捷方式：你可将函数序言替换为指令 enter 0,0，将函数尾声替换为指令 leave。但 enter 的性能很差，因此如果你认为性能是一个问题，则可以继续使用 push rbp 和 mov rbp,rsp。指令 leave 不存在这样的性能问题。

13.3　小结

本章内容：

- 栈对齐
- 使用栈帧
- 使用 SASM 查看栈指针
- enter 和 leave 指令

第 14 章

外部函数

我们已经知道如何在源代码中创建和使用函数。但这些函数不必与主程序保存在同一文件中。可在单独的文件中编写和汇编这些函数，并在编译程序时进行链接。printf 是一个外部函数的例子，前面章节已经使用过多次了。在计划使用外部函数的源文件中，使用关键字 extern 对其进行声明，并且汇编程序知道它不必查找该函数的源。汇编器将假定该函数已在目标文件中进行了汇编。外部函数将由链接器插入，前提是可在目标文件中找到它。

与使用 printf 之类的 C 函数类似，你可以构建自己的函数集并在需要时链接它们。

14.1 编译并链接函数

代码清单 14-1 展示了一个示例程序，其中包含三个源文件，这些文件将另存为单独的文件：function4.asm、circle.asm 和 rect.asm。还有一个新的 makefile。仔细研究一下。

代码清单 14-1：function4.asm

```
; function4.asm
extern printf
extern c_area
extern c_circum
extern r_area
extern r_circum
global pi
section .data
```

```nasm
        pi      dq 3.141592654
        radius  dq 10.0
        side1   dq 4
        side2   dq 5
        fmtf    db "%s %f",10,0
        fmti    db "%s %d",10,0
        ca      db "The circle area is ",0
        cc      db "The circle circumference is ",0
        ra      db "The rectangle area is ",0
        rc      db "The rectangle circumference is ",0
section .bss
section .text
        global main
main:
push rbp
mov rbp,rsp

; 圆面积
    movsd xmm0, qword [radius]      ; 半径,xmm0参数
    call  c_area                    ; 返回的面积存放在xmm0中
    ; 打印圆面积
        mov rdi, fmtf
        mov rsi, ca
        mov rax, 1
        call printf
; 圆周长
    movsd xmm0, qword [radius]      ; 半径,xmm0参数
    call  c_circum                  ; 周长存放在xmm0中
    ; 打印圆周长
        mov rdi, fmtf
        mov rsi, cc
        mov rax, 1
        call printf
; 矩形面积
    mov rdi, [side1]
    mov rsi, [side2]
    call r_area                     ; 返回的面积存放在rax中
    ; 打印矩形面积
        mov rdi, fmti
```

第 14 章 ■ 外部函数

```
            mov  rsi, ra
            mov  rdx, rax
            mov  rax, 0
            call printf
; 矩形周长
        mov  rdi, [side1]
        mov  rsi, [side2]
        call r_circum          ; 周长存放在rax中
        ; 打印矩形周长
            mov rdi, fmti
            mov rsi, rc
            mov rdx, rax
            mov rax, 0
            call printf
mov rsp,rbp
pop rbp
ret
```

在上面的源代码中,我们将许多函数声明为外部函数,就像之前使用 printf 一样,这里没有新内容。但将 pi 声明为全局变量;这意味着外部函数也可使用这个变量。

代码清单 14-2 和代码清单 14-3 展示了仅包含函数的单独文件。

代码清单 14-2:circle.asm

```
; circle.asm
extern pi
section .data
section .bss
section .text
;----------------------------------------------
global c_area
c_area:
        section .text
        push rbp
        mov rbp,rsp
            movsd  xmm1, qword [pi]
            mulsd  xmm0, xmm0      ; 半径存放在xmm0 中
            mulsd  xmm0, xmm1
        mov rsp,rbp
```

```
        pop rbp
        ret
;------------------------------------------------
global c_circum
c_circum:
        section .text
        push rbp
        mov rbp,rsp
            movsd   xmm1, qword [pi]
            addsd   xmm0, xmm0      ;半径存放在 xmm0 中
            mulsd   xmm0, xmm1
        mov rsp,rbp
        pop rbp
        ret
```

代码清单 14-3：rect.asm

```
; rect.asm
section .data
section .bss
section .text
;------------------------------------------------
global r_area
r_area:
        section .text
        push rbp
        mov rbp,rsp
            mov rax, rsi
            imul rax, rdi
            mov rsp,rbp
        pop rbp
        ret
;------------------------------------------------
global r_circum
r_circum:
        section .text
        push rbp
        mov rbp,rsp
            mov rax, rsi
```

```
            add rax, rdi
            add rax, rax
    mov rsp,rbp
    pop rbp
    ret
```

在 circle.asm 中，我们希望把在主源文件中声明的变量 pi 用作全局变量，这不是一个好主意，但此处出于演示目的而这样做。诸如 pi 的全局变量难以跟踪，甚至可能导致名称相同的冲突变量。最佳实践是使用寄存器将值传递给函数。这里，必须指定 pi 是外部变量。circle.asm 和 rect.asm 分别用于计算周长和面积。我们必须指出这些功能是全局的，类似于主程序。汇编这些函数时，将增加必要的"开销"，使链接程序可将这些函数添加到其他目标代码中。

14.2　扩展 makefile

要使所有这些工作正常进行，需要一个扩展的 makefile，如代码清单 14-4 所示。

代码清单 14-4：makefile

```
# function4、circle 和 rect 的 makefile
function4: function4.o circle.o rect.o
    gcc -g -o function4 function4.o circle.o rect.o -no-pie
function4.o: function4.asm
    nasm -f elf64 -g -F dwarf function4.asm -l function4.lst
circle.o: circle.asm
    nasm -f elf64 -g -F dwarf circle.asm -l circle.lst
rect.o: rect.asm
    nasm -f elf64 -g -F dwarf rect.asm -l rect.lst
```

自下而上地阅读 makefile。首先将不同的程序集源文件汇编成对象文件，然后在 function4(可执行文件)中将对象文件链接在一起。可在这里看到使用 make 的力量。当你修改一个源文件时，make 知道哪些文件需要重新汇编和链接，这要归功于树结构。当然，如果你的函数稳定且不再更改，则不必尝试在每个 makefile 中重新汇编它们。只需要将对象文件存储在一个方便的目录中，并在 makefile 的 gcc 行中引用该对象文件及其完整路径。对象文件是汇编或编译源代码的结果。它包含机器代码以及链接器的信息，这些信息说明为了生成有效的可执行文件，需要哪些全局变量和外部函数。在我们的例子中，所有对象文件都与主要源文件位于同一目录中，因此此处未指定任何路径。

那么 printf 函数呢？为什么在 makefile 中没有对 printf 的引用？好吧，gcc 非常聪明，它可以检查 C 库中源代码中引用的函数。这意味着你不应使用 C 函数的名称来命名自己的函数！ 这会让所有人感到困惑，也会导致链接出错。

在代码中，我们使用寄存器将值从主程序传递到函数，反之亦然，这是最佳实践。例如，在调用 r_area 之前，我们将 side1 移至 rdi，将 side2 移至 rsi。然后通过 rax 返回面积(计算结果)。为了返回结果，我们可以使用一个全局变量，类似于 main 的 section .data 中的 pi。但是正如我们之前所说，应该避免这种使用方式。在下一章有关调用约定的内容中，我们将对此进行更详细的介绍。

图 14-1 展示了该程序的输出。

```
jo@UbuntuDesktop:~/Desktop/linux64/gcc/17 function4$ make
nasm -f elf64 -g -F dwarf function4.asm -l function4.lst
nasm -f elf64 -g -F dwarf circle.asm -l circle.lst
nasm -f elf64 -g -F dwarf rect.asm -l rect.lst
gcc -g -o function4 function4.o circle.o rect.o
jo@UbuntuDesktop:~/Desktop/linux64/gcc/17 function4$ ./function4
The circle area is  314.159265
The circle circumference is  62.831853
The rectangle area is  20
The rectangle circumference is  18
jo@UbuntuDesktop:~/Desktop/linux64/gcc/17 function4$
```

图 14-1   function4 的输出

在 SASM 中使用此示例时，必须首先汇编外部函数以获得对象文件。然后，需要在 SASM Settings 对话框的 Build 选项卡上，在 Linking Options 行中添加这些对象文件的位置。这种情况下，Linking Options 行如下所示(请注意不要在此行中引入不必要的空格！)

$PROGRAM.OBJ$ -g -o $PROGRAM$ circle.o rect.o -no-pie

14.3 小结

本章内容：
- 如何使用外部函数
- 如何设置全局变量
- 如何使用 makefile 和外部函数
- 如何在函数之间传递值

第 15 章

调用约定

调用约定描述了如何在函数之间传递变量。如果仅使用自己构建的函数，则不必关心调用约定。但当你使用 C 库中的 C 函数时，你需要知道必须在哪个寄存器中放置该函数要使用的值。另外，如果编写汇编函数来构建供其他开发人员使用的库，则最好遵循一些约定，确定将哪些寄存器用于哪些函数参数。否则，将出现很多参数冲突。

你应该已经注意到，使用 printf 函数时，我们在 rdi 中放入一个参数，在 rsi 中放入了一个参数，并在 xmm0 中放入了另一个参数。我们正在使用的就是一个调用约定。

为了避免冲突和由此导致的崩溃，聪明的开发人员设计了调用约定，这是调用函数的标准化方法。这是一个不错的主意，但如你所料，并非所有人都同意其他人的意见，因此存在几种不同的调用约定。到目前为止，本书中一直使用 System V AMD64 ABI 调用约定，这是 Linux 平台上的标准。还有另一个值得了解的调用约定：用于 Windows 编程的 Microsoft x64 调用约定。

这些调用约定使你不必访问源代码即可使用由汇编语言构建的外部函数以及从 C 之类的语言编译的函数。只需要将正确参数放入调用约定中指定的寄存器中即可。

可在 https://software.intel.com/sites/default/files/article/402129/mpx-linux64-abi.pdf 上找到有关 System V AMD64 ABI 调用约定的更多信息。该英特尔文档包含有关 System V 应用程序二进制接口的大量详细信息。在本章中，我们将展示以标准方式开始调用函数需要了解的内容。

15.1 函数参数

回顾一下以前的源文件：对于圆(circle)的计算，我们使用 xmm0 将浮点值从主程序传递到 circle 函数，然后使用 xmm0 将函数的浮点结果返回给主程序。对于矩形(rectangle)计算，我们使用 rdi 和 rsi 将整数值传递给函数，并将返回结果放入 rax。这种传递参数和结果的方式由调用约定规定。

非浮点参数(例如整数和地址)按以下方式传递：
- 第一个参数放入 rdi。
- 第二个参数放入 rsi。
- 第三个参数放入 rdx。
- 第四个参数放入 rcx。
- 第五个参数放入 r8。
- 第六个参数放入 r9。

其他参数通过栈以相反的顺序传递，这样我们就可以按正确的顺序弹出。例如，有 10 个参数：
- 压入第十个参数。
- 压入第九个参数。
- 压入第八个参数。
- 压入第七个参数。

一旦进入函数，只需要从寄存器中获取值即可。从栈中弹出值时，必须要小心；请记住，在调用函数时，返回地址紧跟在参数之后被压入栈。

- 压入第十个参数时，栈指针 rsp 将减少 8 个字节。
- 压入第九个参数时，rsp 减少 8 个字节。
- 压入第八个参数时，rsp 减少 8 个字节。
- 压入第七个参数时，rsp 减少 8 个字节。
- 调用函数；rip 被压入栈，rsp 减少 8 个字节。
- 在函数开头压入 rbp 作为序言的一部分，rsp 减少 8 个字节。
- 将栈对齐在 16 字节边界上，因此可能需要另一次压入才能将 rsp 减少 8 字节。

因此，在我们压入函数的参数后，至少要在栈上压入另外两个寄存器，即增加 16 个字节。因此，要在函数中访问参数，必须跳过栈上的前 16 个字节，如果必须对齐栈，则可能跳过更多字节。

浮点参数通过 xmm 寄存器传递，如下所示：
- 第一个参数放入 xmm0。

第 15 章 ■ 调用约定

- 第二个参数放入 xmm1。
- 第三个参数放入 xmm2。
- 第四个参数放入 xmm3。
- 第五个参数放入 xmm4。
- 第六个参数放入 xmm5。
- 第七个参数放入 xmm6。
- 第八个参数放入 xmm7。

其他参数通过栈传递。正如你所期望的那样，这不是通过 push 指令完成的。稍后，我们将在更高级的 SIMD 章节中展示如何执行此操作。

函数使用 xmm0 返回浮点结果，并以 rax 返回一个整数或地址。

是不是感觉很复杂？代码清单 15-1 展示了一个使用 printf 打印多个参数的示例。

代码清单 15-1：function5.asm

```
; function5.asm
extern printf
section .data
        first     db    "A",0
        second    db    "B",0
        third     db    "C",0
        fourth    db    "D",0
        fifth     db    "E",0
        sixth     db    "F",0
        seventh   db    "G",0
        eighth    db    "H",0
        ninth     db    "I",0
        tenth     db    "J",0
        fmt1 db   "The string is: %s%s%s%s%s%s%s%s%s%s",10,0
        fmt2      db    "PI = %f",10,0
        pi        dq    3.14
section .bss
section .text
        global main
main:
push  rbp
mov   rbp,rsp

        mov   rdi,fmt1      ;first 使用寄存器
```

101

```
        mov   rsi, first
        mov   rdx, second
        mov   rcx, third
        mov   r8, fourth
        mov   r9, fifth

        push tenth          ; 现在开始按相反的顺序压入
        push ninth
        push eighth
        push seventh
        push sixth
        mov  rax, 0
        call printf
        and  rsp, 0xfffffffffffffff0  ; 16 字节对齐堆栈
        movsd    xmm0,[pi]            ; 现在打印一个浮点数
        mov  rax, 1                   ; 1 个浮点打印
        mov  rdi, fmt2
        call printf
leave
ret
```

在此示例中，我们采用正确顺序将所有参数传递给 printf。注意压入参数的相反顺序。

在调用 printf 前，使用调试器检查 rsp。栈未对齐 16 字节！该程序没有崩溃，因为我们没有要求 printf 打印浮点数。但下一个 printf 需要打印浮点数。在使用 printf 之前，我们必须对齐栈，因此使用以下指令：

```
and rsp, 0xfffffffffffffff0
```

该指令使 rsp 中的所有字节保持不变，除了最后一个字节：rsp 中的最后四位更改为 0，从而减少了 rsp 中的数字并在 16 字节边界上对齐 rsp。如果堆栈已开始对齐，则 and 指令将无效。不过要小心。如果要在此 and 指令之后从堆栈中弹出值，则会遇到一个问题：你必须查明 and 指令是否改变了 rsp，并最终在执行 and 指令之前再次将 rsp 调整为其值。

图 15-1 展示了输出。

```
jo@UbuntuDesktop:~/Desktop/linux64/gcc/18 function5$ make
nasm -f elf64 -g -F dwarf function5.asm -l function5.lst
gcc -g -o function5 function5.o -no-pie
jo@UbuntuDesktop:~/Desktop/linux64/gcc/18 function5$ ./function5
The string is: ABCDEFGHIJ
PI = 3.140000
jo@UbuntuDesktop:~/Desktop/linux64/gcc/18 function5$
```

图 15-1　function5 的输出

15.2　栈布局

下面看一个示例，在此示例中，当压入函数参数时，可以看到栈上发生了什么。代码清单 15-2 展示了一个使用函数构建字符串的程序，当该函数返回时，将打印该字符串。

代码清单 15-2：function6.asm

```
; function6.asm
extern printf
section .data
        first     db    "A"
        second    db    "B"
        third     db    "C"
        fourth    db    "D"
        fifth     db    "E"
        sixth     db    "F"
        seventh   db    "G"
        eighth    db    "H"
        ninth     db    "I"
        tenth     db    "J"
        fmt       db    "The string is: %s",10,0
section .bss
        flist   resb 11      ;字符串的长度+终止 0
section .text
        global main
main:
push rbp
mov    rbp, rsp
        mov   rdi, flist      ; 长度
        mov   rsi, first      ; 放入寄存器
        mov   rdx, second
```

```
            mov     rcx, third
            mov     r8, fourth
            mov     r9, fifth
            push    tenth           ; 以相反的顺序开始压入
            push    ninth
            push    eighth
            push    seventh
            push    sixth
            call    lfunc           ;调用函数
            ; 打印结果
            mov     rdi, fmt
            mov     rsi, flist
            mov     rax, 0
            call    printf
leave
ret
;----------------------------------------------
lfunc:
push    rbp
mov     rbp,rsp
            xor     rax,rax              ;清空 rax（特别是高位字节）
            mov     al,byte[rsi]         ; 将第一个参数移到 al
            mov     [rdi], al            ; 将 al 中的内容存储到内存
            mov     al, byte[rdx]        ; 将第二个参数移到 al
            mov     [rdi+1], al          ; 将 al 中的内容存储到内存
            mov     al, byte[rcx]        ; 同样处理其他参数
            mov     [rdi+2], al
            mov     al, byte[r8]
            mov     [rdi+3], al
            mov     al, byte[r9]
            mov     [rdi+4], al
; 从堆栈获取参数
            push    rbx                         ; 被调用者保留
            xor     rbx,rbx
            mov     rax, qword [rbp+16]  ; 初始化堆栈+rip+rbp
            mov     bl, byte[rax]        ; 提取字符
            mov     [rdi+5], bl          ; 把字符存储到内存
            mov     rax, qword [rbp+24]  ; 继续下一个值
            mov     bl, byte[rax]
```

```
            mov     [rdi+6], bl
            mov     rax, qword [rbp+32]
            mov     bl, byte[rax]
            mov     [rdi+7], bl
            mov     rax, qword [rbp+40]
            mov     bl, byte[rax]
            mov     [rdi+8], bl
            mov     rax, qword [rbp+48]
            mov     bl, byte[rax]
            mov     [rdi+9], bl
            mov     bl,0
            mov     [rdi+10], bl
    pop rbx              ; 被调用者保留
    mov rsp,rbp
    pop rbp
    ret
```

这里，没有像上一节中那样在提供所有参数之后立即使用 printf 进行打印，而是调用了 lfunc 函数。这个函数接收所有参数，并在内存中建立一个字符串(flist)；返回 main 后，该字符串将被打印。

查看 lfunc 函数。我们使用如下指令仅占用参数寄存器的低字节，即字符所在的位置：

```
mov al,byte[rsi]
```

我们将这些字符从 rdi 的地址(即 flist 的地址)开始，逐个存储在内存中，指令如下：mov [rdi], al。使用 byte 关键字是不必要的，但是它可以提高代码的可读性。

当开始从栈中弹出值时，会变得很有趣。在 lfunc 的开头，将 rsp 的值(即栈地址)保存到 rbp 中。但在执行这条指令以及将值压入 main 的过程中，rsp 被修改了两次。首先，当调用 lfunc 时，返回地址被压入栈。然后将 rbp 作为序言的一部分。总计，rsp 减少了 16 个字节。要访问压入的值，必须将地址的值增加 16 个字节。这就是为什么用它来访问变量 sixth：

```
mov rax, qword [rbp+16]
```

其他变量都比前一个高 8 字节。我们使用 rbx 作为在 flist 中构建字符串的临时寄存器。在使用 rbx 之前，将 rbx 的内容保存到栈中。你永远不知道 rbx 是否在 main 中用于其他目的，因此保留 rbx 并在离开函数前将其还原。

图 15-2 展示了输出。

```
jo@UbuntuDesktop:~/Desktop/linux64/gcc/18 function6$ make
nasm -f elf64 -g -F dwarf function6.asm -l function6.lst
gcc -g -o function6 function6.o -no-pie
jo@UbuntuDesktop:~/Desktop/linux64/gcc/18 function6$ ./function6
The string is: ABCDEFGHIJ
jo@UbuntuDesktop:~/Desktop/linux64/gcc/18 function6$
```

图 15-2 function6 的输出

15.3 保留寄存器

下面将解释这些指令。

```
push    rbx     ; callee saved
```

和

```
pop     rbx     ; callee saved
```

应当清楚的是，你必须在函数调用期间跟踪寄存器发生的情况。在执行函数期间，某些寄存器将被更改，而某些寄存器则保持不变。你需要采取预防措施，以避免因函数修改你在主调用程序中使用的寄存器而导致意外结果。

表 15-1 概述了调用约定中指定的内容。

表 15-1 调用约定

寄存器	用例	保存
rax	返回值	调用者(Caller)
rbx	被调用者保存	被调用者(Callee)
rcx	第四个参数	调用者(Caller)
rdx	第三个参数	调用者(Caller)
rsi	第二个参数	调用者(Caller)
rdi	第一个参数	调用者(Caller)
rbp	被调用者保存	被调用者(Callee)
rsp	栈指针	被调用者(Callee)
r8	第五个参数	调用者(Caller)
r9	第六个参数	调用者(Caller)
r10	临时	调用者(Caller)
r11	临时	调用者(Caller)
r12	被调用者保存	被调用者(Callee)

(续表)

寄存器	用例	保存
r13	被调用者保存	被调用者(Callee)
r14	被调用者保存	被调用者(Callee)
r15	被调用者保存	被调用者(Callee)
xmm0	第一个参数和返回值	调用者(Caller)
xmm1	第二个参数和返回值	调用者(Caller)
xmm2-7	参数	调用者(Caller)
xmm8-15	临时	调用者(Caller)

被调用的函数就是被调用者。当函数使用被调用者保存的寄存器时，函数需要在使用之前将该寄存器压入栈，然后按正确顺序弹出它。调用者希望被调用者保存的寄存器在函数调用后保持不变。参数寄存器可在函数执行期间更改，因此如果必须保留它们，则调用者有责任压入/弹出它们。同样，可在函数中更改临时寄存器，因此如有必要，调用者需要将其压入/弹出。返回值 rax 需要由调用者压入/弹出。

修改现有函数并开始使用保存调用者的寄存器时，可能会突然出现问题。如果你没有在调用者中添加该寄存器的压入/弹出操作，则会产生意外结果。

被调用者保存的寄存器也称为非易失性寄存器。调用者必须保存的寄存器也称为易失性(volatile)寄存器。

xmm 寄存器都可以通过一个函数进行更改。如有必要，调用者将负责保存它们。

当然，如果确定不使用更改的寄存器，则可以不保存这些寄存器。但是，如果将来更改代码，但开始使用这些寄存器未保存它们，则可能遇到麻烦。信不信由你，几周或几个月后，汇编代码很难阅读，即使你自己编写了所有代码也是如此。

最后注意，syscall 也是一个函数，它将修改寄存器，因此请密切注意 syscall 在做什么。

15.4 小结

本章内容：
- 调用约定
- 栈对齐
- 被调用者/调用者保存的寄存器

第 16 章

位运算

我们已经在第 9 章中使用了整数运算的位运算：移位运算的 sar 和 sal 是向右或向左移位。同样，上一章介绍的用于栈对齐的 and 指令也是位运算。

16.1 基础

在下面的示例程序中，构建一个名为 printb 的自定义 C 函数，以打印一个位字符串。为方便起见，将 64 位字符串分成 8 个字节，每个 8 位。作为练习，完成本章后，请看一下 C 代码，你应该能够编写一个汇编程序来构建一个位字符串。

代码清单 16-1、代码清单 16-2 和代码清单 16-3 分别显示了汇编、C printb 程序和 makefile 中位运算的示例代码。

代码清单 16-1：bits1.asm

```
; bits1.asm
extern printb
extern printf
section .data
    msgn1   db    "Number 1",10,0
    msgn2   db    "Number 2",10,0
    msg1    db    "XOR",10,0
    msg2    db    "OR",10,0
    msg3    db    "AND",10,0
    msg4    db    "NOT number 1",10,0
    msg5    db    "SHL 2 lower byte of number 1",10,0
    msg6    db    "SHR 2 lower byte of number 1",10,0
    msg7    db    "SAL 2 lower byte of number 1",10,0
```

```
        msg8    db      "SAR 2 lower byte of number 1",10,0
        msg9    db      "ROL 2 lower byte of number 1",10,0
        msg10   db      "ROL 2 lower byte of number 2",10,0
        msg11   db      "ROR 2 lower byte of number 1",10,0
        msg12   db      "ROR 2 lower byte of number 2",10,0
        number1         dq      -72
        number2         dq      1064
section .bss
section .text
        global main
main:
push rbp
mov rbp,rsp
; 打印 number1
        mov rsi, msgn1
        call printmsg
        mov rdi, [number1]
        call printb
; 打印 number2
        mov rsi, msgn2
        call printmsg
        mov rdi, [number2]
        call printb

; 打印 XOR(exclusive OR,异或)------------------------
        mov rsi, msg1
        call printmsg
; 进行 xor 运算并打印
        mov rax,[number1]
        xor  rax,[number2]
        mov rdi, rax
        call printb

; 打印 OR(或) ----------------------------------------
        mov rsi, msg2
        call printmsg
; 进行 or 运算并打印
        mov rax,[number1]
        or   rax,[number2]
        mov rdi, rax
```

```
        call printb

; 打印 AND(和) -------------------------------------
        mov rsi, msg3
        call printmsg
; 进行 and 运算并打印
        mov rax,[number1]
        and rax,[number2]
        mov rdi, rax
        call printb

; 打印 NOT(非) -------------------------------------
        mov rsi, msg4
        call printmsg
; 进行 not 运算并打印
        mov rax,[number1]
        not   rax
        mov rdi, rax
        call printb

; 打印 SHL(shift left,左移----------------------------
        mov rsi, msg5
        call printmsg
; 进行 shl 运算并打印
        mov rax,[number1]
        shl   al,2
        mov rdi, rax
        call printb

; 打印 SHR(shift right,右移)--------------------------
        mov rsi, msg6
        call printmsg
;进行 shr 运算并打印
        mov rax,[number1]
        shr   al,2
        mov rdi, rax
        call printb

; 打印 SAL(shift arithmetic left,左移算术)----------------
        mov rsi, msg7
        call printmsg
```

```
        ; 进行 sal 运算并打印
                mov rax,[number1]
                sal  al,2
                mov rdi, rax
                call printb

        ; 打印 SAR(shift arithmetic right，右移算术)-----------------
                mov rsi, msg8
                call printmsg
        ; 进行 sar 运算并打印
                mov rax,[number1]
                sar  al,2
                mov rdi, rax
                call printb

        ; 打印 ROL(rotate left，向左旋转)---------------------------
                mov rsi, msg9
                call printmsg
        ; 进行 rol 运算并打印
                mov rax,[number1]
                rol  al,2
                mov rdi, rax
                call printb
                mov rsi, msg10
                call printmsg
                mov rax,[number2]
                rol  al,2
                mov rdi, rax
                call printb

        ; 打印 ROR(rotate right，向右旋转)---------------------------
                mov rsi, msg11
                call printmsg
        ; 进行 ror 运算并打印
                mov rax,[number1]
                ror  al,2
                mov rdi, rax
                call printb
                mov rsi, msg12
                call printmsg
```

```
            mov rax,[number2]
            ror   al,2
            mov rdi, rax
            call printb
leave
ret

;----------------------------------------------------
printmsg:         ; 打印每个位操作的标题
section .data
        .fmtstr   db "%s",0
section .text
        mov rdi,.fmtstr
        mov rax,0
        call printf
        ret
```

代码清单 16-2：printb.c

```c
// printb.c

#include <stdio.h>

void printb(long long n){
    long long s,c;
    for (c = 63; c >= 0; c--)
    {
    s = n >> c;
    // 每 8 位(bit)后的空间
    if ((c+1) % 8 == 0) printf(" ");

    if (s & 1)
        printf("1");
    else
        printf("0");
    }
    printf("\n");
}
```

代码清单 16-3：bit1 和 printb 的 makefile

```
# bits1 和 printb 的 makefile
bits1: bits1.o printb.o
      gcc -g -o bits1 bits1.o printb.o -no-pie
bits1.o: bits1.asm
      nasm -f elf64 -g -F dwarf bits1.asm -l bits1.lst
printb: printb.c
      gcc -c printb.c
```

编译并运行程序，然后研究一下输出。如果使用的是 SASM，请不要忘记先编译 printb.c 文件，然后在链接选项中添加对象文件，如第 14 章中讨论外部函数时所述。

这是一个很长的程序。幸运的是，代码并不复杂。我们将展示不同位运算的工作方式。使用图 16-1 所示的输出来帮助你完成代码。

```
jo@UbuntuDesktop:~/Desktop/linux64/gcc/20_bits1$ make
nasm -f elf64 -g -F dwarf bits1.asm -l bits1.lst
cc    -c -o printb.o printb.c
gcc -g -o bits1 bits1.o printb.o
jo@UbuntuDesktop:~/Desktop/linux64/gcc/20_bits1$ ./bits1
Number 1
 11111111 11111111 11111111 11111111 11111111 11111111 11111111 10111000
Number 2
 00000000 00000000 00000000 00000000 00000000 00000000 00000100 00101000
XOR
 11111111 11111111 11111111 11111111 11111111 11111111 11111011 10010000
OR
 11111111 11111111 11111111 11111111 11111111 11111111 11111111 10111000
AND
 00000000 00000000 00000000 00000000 00000000 00000000 00000100 00101000
NOT number 1
 00000000 00000000 00000000 00000000 00000000 00000000 00000000 01000111
SHL 2 lower byte of number 1
 11111111 11111111 11111111 11111111 11111111 11111111 11111111 11100000
SHR 2 lower byte of number 1
 11111111 11111111 11111111 11111111 11111111 11111111 11111111 00101110
SAL 2 lower byte of number 1
 11111111 11111111 11111111 11111111 11111111 11111111 11111111 11100000
SAR 2 lower byte of number 1
 11111111 11111111 11111111 11111111 11111111 11111111 11111111 11101110
ROL 2 lower byte of number 1
 11111111 11111111 11111111 11111111 11111111 11111111 11111111 11100010
ROL 2 lower byte of number 2
 00000000 00000000 00000000 00000000 00000000 00000000 00000100 10100000
ROR 2 lower byte of number 1
 11111111 11111111 11111111 11111111 11111111 11111111 11111111 00101110
ROR 2 lower byte of number 2
 00000000 00000000 00000000 00000000 00000000 00000000 00000100 00001010
jo@UbuntuDesktop:~/Desktop/linux64/gcc/20_bits1$
```

图 16-1　bits1.asm 的输出

首先请注意 number1(-72)的二进制表示形式；最高有效位中的 1 表示负数。

指令 xor、or、and 和 not 很简单，它们的工作原理如第 5 章所述。用不同的值进行实验，看看它是如何工作的。对于 shl、shr、sal 和 sar，我们使用 rax 的低位字节来说明发生了什么。在 shl 中，向左移位，零加在 al 的右边；向左移位时，从第 8 位向左移的位被丢弃。使用 shr，位将向右移，而在 al 的左侧用 0 补齐。所有位都向右移动，最低有效位中向右移动的位被丢弃。当你单步执行程序时，请注意标志寄存器，尤其是符号寄存器和溢出寄存器。

算术左移 sal 与 shl 完全相同；它将值相乘。算术右移 sar 与 shr 不同，使用除法。这就是所谓的符号扩展。如果 al 中最左边的位是 1，则 al 包含的是负值。为了正确地进行运算，右移时如果值为负值，则在左边加 1 而不是 0。这称为符号扩展。

向左旋转 rol，删除最左边的位，向左移动，然后将删除的位添加到右边。向右旋转 ror 以类似的方式工作。

16.2 算术

下面深入研究一下移位算法。为什么有两种左移和两种右移呢？当用负值进行算术运算时，移位指令可能给出错误结果，因为需要考虑符号扩展。这就是为什么有算术移位指令和逻辑移位指令。

研究代码清单 16-4 中的示例。

代码清单 16-4：bits2.asm

```
; bits2.asm
extern printf
section .data
    msgn1   db   "Number 1 is = %d",0
    msgn2   db   "Number 2 is = %d",0
    msg1    db   "SHL 2 = OK multiply by 4",0
    msg2    db   "SHR 2 = WRONG divide by 4",0
    msg3    db   "SAL 2 = correctly multiply by 4",0
    msg4    db   "SAR 2 = correctly divide by 4",0
    msg5    db   "SHR 2 = OK divide by 4",0

    number1  dq   8
    number2  dq   -8
    result   dq   0

section .bss
```

```
section .text
        global main
main:
push rbp
mov rbp,rsp

;SHL-------------------------------------------------

;正数
        mov   rsi, msg1
        call  printmsg      ;打印标题
        mov   rsi, [number1]
        call  printnbr      ;打印 number1
        mov   rax,[number1]
        shl   rax,2         ;乘以 4(逻辑)
        mov   rsi, rax
        call  printres
;负数
        mov   rsi, msg1
        call  printmsg      ;打印标题
        mov   rsi, [number2]
        call  printnbr      ;打印 number2
        mov   rax,[number2]
        shl   rax,2         ;乘以 4(逻辑)
        mov   rsi, rax
        call  printres
;SAL-------------------------------------------------
;正数
        mov   rsi, msg3
        call  printmsg      ;打印标题
        mov   rsi, [number1]
        call  printnbr      ;打印 number1
        mov   rax,[number1]
        sal   rax,2         ;乘以 4(算术)
        mov   rsi, rax
        call  printres
;负数
        mov   rsi, msg3
        call  printmsg      ;打印标题
        mov   rsi, [number2]
```

```
        call printnbr      ;打印 number2
        mov  rax,[number2]
        sal  rax,2         ;乘以 4(算术)
        mov  rsi, rax
        call printres
;SHR----------------------------------------------
;正数
        mov rsi, msg5
        call printmsg      ;打印标题
        mov  rsi, [number1]
        call printnbr      ;打印 number1
        mov  rax,[number1]
        shr  rax,2         ;除以 4 (逻辑)
        mov  rsi, rax
        call printres
;负数
        mov  rsi, msg2
        call printmsg      ;打印标题
        mov  rsi, [number2]
        call printnbr      ;打印 number2
        mov  rax,[number2]
        shr  rax,2         ;除以 4 (逻辑)
        mov  [result], rax
        mov  rsi, rax
        call printres
;SAR----------------------------------------------
;正数
        mov  rsi, msg4
        call printmsg      ;打印标题
        mov  rsi, [number1]
        call printnbr      ;打印 number1
        mov  rax,[number1]
        sar  rax,2         ;除以 4 (算术)
        mov  rsi, rax
        call printres
;负数
        mov  rsi, msg4
        call printmsg      ;打印标题
        mov  rsi, [number2]
```

```
        call printnbr        ;打印number2
        mov  rax,[number2]
        sar  rax,2           ;除以4（算术）
        mov  rsi, rax
        call printres
leave
ret
;----------------------------------
printmsg:           ;打印标题
        section .data
            .fmtstr db 10,"%s",10,0   ;格式化字符串
        section .text
            mov  rdi,.fmtstr
            mov  rax,0
            call printf
        ret
;----------------------------------
printnbr:           ;打印数字
        section .data
            .fmtstr db "The original number is %lld",10,0
        section .text
            mov  rdi,.fmtstr
            mov  rax,0
            call printf
        ret
;----------------------------------
printres:           ;打印结果
section .data
    .fmtstr db "The resulting number is %lld",10,0
section .text
    mov  rdi, .fmtstr
    mov  rax,0
    call printf
ret
```

使用图 16-2 所示的输出来分析代码。

```
jo@UbuntuDesktop:~/Desktop/linux64/gcc/21 bits2$ make
nasm -f elf64 -g -F dwarf bits2.asm -l bits2.lst
gcc -g -o bits2 bits2.o
jo@UbuntuDesktop:~/Desktop/linux64/gcc/21 bits2$ ./bits2
SHL 2 = OK multiply by 4
The original number is 8
The resulting number is 32

SHL 2 = OK multiply by 4
The original number is -8
The resulting number is -32

SAL 2 = correctly multiply by 4
The original number is 8
The resulting number is 32

SAL 2 = correctly multiply by 4
The original number is -8
The resulting number is -32

SHR 2 = OK divide by 4
The original number is 8
The resulting number is 2

SHR 2 = wrong divide by 4
The original number is -8
The resulting number is 4611686018427387902

SAR 2 = correctly divide by 4
The original number is 8
The resulting number is 2

SAR 2 = correctly divide by 4
The original number is -8
The resulting number is -2
jo@UbuntuDesktop:~/Desktop/linux64/gcc/21 bits2$
```

图 16-2　bits2.asm 的输出

注意，shl 和 sal 给出了相同的结果，也有负数。但是要小心，如果 shl 在最左边的位上加 1 而不是 0，结果就会变成负数，并且是错误的。

只有当数字为正时，shr 和 sar 指令才给出相同的结果。当使用带负数的 shr 时，算术结果是错误的，这是因为 shr 没有符号扩展。

因此可以得出结论：在执行算术运算时，使用 sal 和 sar。

有简单直接的指令(例如乘法和除法)，为什么要使用移位呢？事实证明，移位比乘法或除法指令快得多。通常，位指令非常快；例如 xor rax, rax 比 mov rax, 0 快。

16.3　小结

本章内容：
- 用于位运算的汇编指令
- 逻辑和算术移位的区别

第 17 章

位操作

你已经知道可以使用位运算(例如 and、xor 或 or 和 not)来设置或清除位。但是还有其他修改单个位的方法：bts 用于将位设置为 1，btr 用于将位设置为 0，bt 用于测试位是否设置为 1。

17.1 修改位的其他方法

代码清单 17-1 展示了示例代码。

代码清单 17-1：bits3.asm

```
; bits3.asm
extern printb
extern printf
section .data
    msg1    db  "No bits are set:",10,0
    msg2    db  10,"Set bit #4, that is the 5th bit:",10,0
    msg3    db  10,"Set bit #7, that is the 8th bit:",10,0
    msg4    db  10,"Set bit #8, that is the 9th bit:",10,0
    msg5    db  10,"Set bit #61, that is the 62nd bit:",10,0
    msg6    db  10,"Clear bit #8, that is the 9th bit:",10,0
    msg7    db  10,"Test bit #61, and display rdi",10,0
    bitflags dq 0
section .bss
section .text
    global main
main:
```

```asm
        push rbp
        mov rbp,rsp
                ;打印标题
                mov rdi, msg1
                xor  rax,rax
                call printf

                ;打印bitflags
                mov rdi, [bitflags]
                call printb

;设置bit 4 (第5个位)
                ;打印标题
                mov  rdi, msg2
                xor  rax,rax
                call printf

                bts  qword [bitflags],4     ; 设置bit 4
                ;打印bitflags
                mov  rdi, [bitflags]
                call printb

;设置bit 7 (第8个位)
                ;打印标题
                mov rdi, msg3
                xor rax,rax
                call printf

                bts  qword [bitflags],7     ; 设置bit 7
                ;打印bitflags
                mov rdi, [bitflags]
                call printb

;设置bit 8 (第9个位)
                ;打印标题
                mov rdi, msg4
                xor  rax,rax
                call printf

                bts  qword [bitflags],8     ; 设置bit 8
                ;打印bitflags
                mov rdi, [bitflags]
```

```
            call printb
;设置 bit 61（第 62 个位）
        ;打印标题
        mov rdi, msg5
        xor  rax,rax
        call printf

        bts  qword [bitflags],61   ; 设置 bit 61
        ;打印 bitflags
        mov rdi, [bitflags]
        call printb

;清除 bit 8（第 9 个位）
        ;打印标题
        mov rdi, msg6
        xor rax, rax
        call printf

        btr  qword [bitflags],8  ; 重置 bit 8
        ;打印 bitflags
        mov rdi, [bitflags]
        call printb

; 测试 bit 61（如果为 1，将设置进位标志 CF）
        ;打印标题
        mov  rdi, msg7
        xor  rax, rax
        call printf
        xor  rdi,rdi
        mov  rax,61          ; bit 61 被测试
        xor  rdi,rdi         ; 确保所有位都为 0
        bt   [bitflags],rax  ; 测试位
        setc dil             ; 如果设置了 CF，则将 dil(rdi 的低位)设置为 1
        call printb          ; 显示 rdi
leave
ret
```

这里再次使用 printb.c 程序，确保相应调整你的 makefile 或 SASM 编译设置。变量位标志(bitflags)是这里的研究对象。我们将操作这个变量中的位。

17.2 位标志变量

请记住，位计数(索引)从 0 开始。这意味着在一个具有 8 位的字节中，第一个位的索引为 0，最后一位的索引为 7。使用 bts 指令将位设置为 1，并使用 btr 将位重置为 0 很简单：只需要将要更改的位的索引指定为第二个操作数。

测试有点复杂。将要测试的位的索引放在 rax 中，并使用指令 bt。如果该位为 1，则进位标志 CF 将被设置为 1；否则，CF 将为 0。根据该标志的值，你可以指示程序是否执行某些指令。这种情况下，我们使用一个特殊指令 setc，这是一个条件集。这种情况下，如果进位标志为 1，则指令将 dil 设置为 1。dil 是 rdi 的低位；在使用 setc 设置 dil 之前，请小心地将 rdi 设置为 0。rdx 的高位很可能是在执行前一条指令时设置的。

setc 指令是 setCC 的一个例子。如果满足 CC 中的条件，则 setCC 在操作数中设置一个字节，其中 CC 是一个标志，例如 CF(缩写为 c)、ZF CF(缩写为 z)、SF CF(缩写为 s)等。要了解更多信息，请参阅英特尔手册。

图 17-1 展示了程序的输出。

```
jo@UbuntuDesktop:~/Desktop/linux64/gcc/22 bits3$ make
nasm -f elf64 -g -F dwarf bits3.asm -l bits3.lst
cc     -c -o printb.o printb.c
gcc -g -o bits3 bits3.o printb.o
jo@UbuntuDesktop:~/Desktop/linux64/gcc/22 bits3$ ./bits3
No bits are set:
 00000000 00000000 00000000 00000000 00000000 00000000 00000000 00000000

Set bit #4, that is the 5th bit:
 00000000 00000000 00000000 00000000 00000000 00000000 00000000 00010000

Set bit #7, that is the 8th bit:
 00000000 00000000 00000000 00000000 00000000 00000000 00000000 10010000

Set bit #8, that is the 9th bit:
 00000000 00000000 00000000 00000000 00000000 00000000 00000001 10010000

Set bit #61, that is the 62nd bit:
 00100000 00000000 00000000 00000000 00000000 00000000 00000001 10010000

Clear bit #8, that is the 9th bit:
 00100000 00000000 00000000 00000000 00000000 00000000 00000000 10010000

Test bit #61, and display dl
 00000000 00000000 00000000 00000000 00000000 00000000 00000000 00000001
jo@UbuntuDesktop:~/Desktop/linux64/gcc/22 bits3$
```

图 17-1　bits3.asm 的输出

17.3 小结

本章内容：
- 使用 btr、bts、bt 设置位、重置位和检查位
- setCC 指令

第 18 章

宏

在程序中多次使用同一组指令时,可创建一个函数,并在每次需要执行指令时调用该函数。但是,函数会降低性能;每次调用函数时,执行都会跳转到内存中某个位置的函数,完成后会跳转回调用程序。从函数调用和返回都需要时间。

为避免这种性能问题,可以使用宏(Macro)。与函数相似,宏是一系列指令。可以给宏分配一个名称,当需要在代码中执行该宏时,只需要指定宏名称,并最终附带参数即可。

区别在于:进行汇编时,在代码中任何"调用"宏的地方,NASM 都会用宏定义中的指令替换宏名称。在执行时,没有来回跳转;NASM 已经在需要的位置插入了机器代码。

宏不是 Intel 汇编语言中的功能,而是 NASM(或另一版本的汇编器)提供的功能。宏是使用预处理器指令创建的,而 NASM 使用宏处理器将宏转换为机器语言,并将机器语言插入代码中的适当位置。

宏将提高代码的执行速度,但也会增加代码的大小,因为在汇编时,宏中的指令会插入你使用宏的每个位置。

有关 NASM 宏的更多信息,请参阅 NASM 手册第 4 章 "NASM 预处理器"(适用于 NASM 版本 2.14.02)。

18.1 编写宏

代码清单 18-1 展示了宏的一些示例。

代码清单 18-1：macro.asm

```nasm
; macro.asm
extern printf

%define    double_it(r)   sal r, 1           ; 单行宏

%macro     prntf 2                           ; 带有两个参数的多行宏
    section .data
        %%arg1    db %1,0                    ; 第一个参数
        %%fmtint  db "%s %ld",10,0           ; 格式化字符串
    section .text                            ; printf 参数
        mov    rdi,%%fmtint
        mov    rsi,%%arg1
        mov    rdx,[%2]                      ; 第二个参数
        mov    rax,0                         ; 没有浮点数
        call   printf
%endmacro

section .data
    number dq 15
section .bss
section .text
    global main
main:
push rbp
mov rbp,rsp
    prntf    "The number is", number
    mov      rax, [number]
    double_it(rax)
    mov      [number],rax
    prntf    "The number times 2 is", number
leave
ret
```

宏有两种：单行宏和多行宏。单行宏以%define 开头。关键字%macro 和%endmacro 之间包含多行宏。关键字%define、%macro 和%endmacro 称为汇编程序预处理器指令。

单行宏非常简单：在汇编时，将指令 double_it(rax)替换为 sal r, 1 的机器代码，其中 r 是 rax 中的值。

多行宏有些复杂。prntf 有两个参数。可以看到，在宏定义中，prntf 后跟数字 2 以指示参数的数量。要在宏内部使用参数，第一个参数用%1 表示，第二个参数用 %2 表示，以此类推。请注意如何通过%1 来使用字符串，以及如何通过[%2](带方括号)来使用数值，这与不使用宏时所需的类似。

可以在宏内使用变量，最好在名称前加上%%，如%%arg1 和%%fmtint。如果省略%%，NASM 会在 prntf 的第一次调用中创建宏变量，但在 prntf 的第二次调用时会引发汇编错误——试图重新定义 arg1 和 fmtint。%%告诉 NASM 为每次宏调用创建变量的新实例(练习：删除%%并尝试汇编)。

汇编宏有一个大问题：它们使调试复杂化！尝试使用 GDB 或基于 GDB 的调试器(如 SASM)调试程序，以查看行为。

图 18-1 展示了输出。

```
jo@UbuntuDesktop:~/Desktop/linux64/gcc/23 macro$ make
nasm -f elf64 -g -F dwarf macro.asm -l macro.lst
gcc -o macro macro.o
jo@UbuntuDesktop:~/Desktop/linux64/gcc/23 macro$ ./macro
The number is 15
The number times 2 is 30
jo@UbuntuDesktop:~/Desktop/linux64/gcc/23 macro$
```

图 18-1　macro.asm 的输出

18.2　使用 objdump

下面验证一下，每次使用宏时，会将汇编宏代码插入可执行文件中的适当位置。为此，我们将使用称为 objdump 的命令行工具。如果按照本书开始时的建议安装了开发工具，则 objdump 已安装。在命令行上输入以下内容：

```
objdump -M intel -d macro
```

标志-M intel 将为我们提供 Intel 语法的代码，而-d macro 将反汇编我们的宏可执行文件。向<main>部分滚动代码。

如图 18-2 所示，prntf 的代码从内存地址 4004f4 插入 400515，从 40052d 插入 40054e。double_it 的代码在地址 400522。汇编器可以自由地将 sal 指令更改为 shl，这是出于性能方面的考虑。正如你在第 16 章有关移位指令的内容中所记得的那样，大多数情况下，这样做可能毫无问题。在执行此操作时，请将 sal 指令更改为 sar。你将看到汇编器不会将 sar 更改为 shr，从而避免了问题。

命令行工具 objdump 可用于查看代码，甚至包括你自己未编写的代码。你可使用 objdump 找到许多有关可执行文件的信息，但本书中将不做详细介绍。如果想了解更多信息，请在命令行输入 man objdump 或在互联网上搜索一下。

```
00000000004004f0 <main>:
  4004f0:	55                   	push   rbp
  4004f1:	48 89 e5             	mov    rbp,rsp
  4004f4:	48 bf 46 10 60 00 00 	movabs rdi,0x601046
  4004fb:	00 00 00
  4004fe:	48 be 38 10 60 00 00 	movabs rsi,0x601038
  400505:	00 00 00
  400508:	48 8b 14 25 30 10 60 	mov    rdx,QWORD PTR ds:0x601030
  40050f:	00
  400510:	b8 00 00 00 00       	mov    eax,0x0
  400515:	e8 d6 fe ff ff       	call   4003f0 <printf@plt>
  40051a:	48 8b 04 25 30 10 60 	mov    rax,QWORD PTR ds:0x601030
  400521:	00
  400522:	48 d1 e0             	shl    rax,1
  400525:	48 89 04 25 30 10 60 	mov    QWORD PTR ds:0x601030,rax
  40052c:	00
  40052d:	48 bf 64 10 60 00 00 	movabs rdi,0x601064
  400534:	00 00 00
  400537:	48 be 4e 10 60 00 00 	movabs rsi,0x60104e
  40053e:	00 00 00
  400541:	48 8b 14 25 30 10 60 	mov    rdx,QWORD PTR ds:0x601030
  400548:	00
  400549:	b8 00 00 00 00       	mov    eax,0x0
  40054e:	e8 9d fe ff ff       	call   4003f0 <printf@plt>
  400553:	c9                   	leave
  400554:	c3                   	ret
  400555:	66 2e 0f 1f 84 00 00 	nop    WORD PTR cs:[rax+rax*1+0x0]
  40055c:	00 00 00
  40055f:	90                   	nop
```

图 18-2　objdump -M intel -d macro

18.3　小结

本章内容：

- 何时使用宏以及何时使用函数
- 单行宏
- 多行宏
- 将参数传递给多行宏
- GDB 的汇编宏问题
- objdump

第 19 章

控制台 I/O

我们已经知道如何使用系统调用或 printf 进行控制台输出。本章将再次使用系统调用，不仅在屏幕上显示，而且从键盘接收输入。

19.1 使用 I/O

可以轻松地从 C 库中借用函数，但这会破坏汇编的乐趣！因此，看看代码清单 19-1 所展示的示例源代码。

代码清单 19-1：console1.asm

```
; console1.asm
section .data
      msg1      db    "Hello, World!",10,0
      msg1len   equ   $-msg1
      msg2      db    "Your turn: ",0
      msg2len   equ   $-msg2
      msg3      db    "You answered: ",0
      msg3len   equ   $-msg3
      inputlen equ    10       ;inputbuffer 的长度
section .bss
      input   resb inputlen+1         ;为终止 0 提供空间
section .text
      global main
main:
push rbp
mov rbp,rsp
```

```
        mov   rsi, msg1        ; 打印第一个字符串
        mov   rdx, msg1len
        call  prints
        mov   rsi, msg2        ; 打印第二个字符串,没有换行
        mov   rdx, msg2len
        call  prints
        mov   rsi, input       ; inputbuffer 的地址
        mov   rdx, inputlen    ; inputbuffer 的长度
        call  reads            ; 等待输入
        mov   rsi, msg3        ; 打印第三个字符串
        mov   rdx, msg3len
        call  prints
        mov   rsi, input       ; 打印 inputbuffer
        mov   rdx, inputlen    ; inputbuffer 的长度
        call  prints
leave
ret
;-------------------------------------------------------
prints:
push rbp
mov rbp, rsp
; rsi 包含字符串的地址
; rdx 包含字符串的长度
        mov rax, 1             ; 1 表示写入
        mov rdi, 1             ; 1 表示标准输出
        syscall
leave
ret
;-------------------------------------------------------
reads:
push rbp
mov rbp, rsp
; rsi 包含 inputbuffer 的地址
; rdi 包含 inputbuffer 的长度
        mov rax, 0             ; 0 表示读取
        mov rdi, 1             ; 表示标准输入
        syscall
leave
ret
```

这并不复杂：我们提供了一个名为 input 的输入缓冲区，用于存储输入中的字符。我们还使用 inputlen 指定了缓冲区的长度。显示一些欢迎消息后，我们将调用 reads 函数，该函数接收通过键盘输入的所有字符，并在按下回车键时将其返回给调用者。然后，调用程序使用 prints 函数显示输入的字符。输出如图 19-1 所示。

```
jo@ubuntu18:~/Desktop/Book/24 console 1$ make
nasm -f elf64 -g -F dwarf console1.asm -l console1.lst
gcc -o console1 console1.o -no-pie
jo@ubuntu18:~/Desktop/Book/24 console 1$ ./console1
Hello, World!
Your turn: Hi There!
You answered: Hi There!
jo@ubuntu18:~/Desktop/Book/24 console 1$
```

图 19-1　console1.asm 的输出

但也有一些问题，我们为输入缓冲区保留了 10 个字节，如果输入超过 10 个字符会怎样？图 19-2 显示了结果。

```
jo@ubuntu18:~/Desktop/Book/24 console 1$ ./console1
Hello, World!
Your turn: Hi there, how are you?
You answered: Hi there, jo@ubuntu18:~/Desktop/Book/24 console 1$ how are you?

Command 'how' not found, did you mean:

  command 'show' from deb mailutils-mh
  command 'show' from deb mmh
  command 'show' from deb nmh
  command 'cow' from deb fl-cow
  command 'hoz' from deb hoz
  command 'sow' from deb ruby-hoe
  command 'hot' from deb hopenpgp-tools

Try: sudo apt install <deb name>

jo@ubuntu18:~/Desktop/Book/24 console 1$
```

图 19-2　多字符情况下 console1.asm 的输出

该程序仅接收十个字符，并且不知道该如何处理多余的字符，因此会将其返回给操作系统。操作系统尝试找出字符并将其解释为 CLI 命令，但由于找不到相应的命令，所有会报错。

这种处理输入的方式可能导致安全漏洞，黑客可在其中破坏程序并获得对操作系统的访问权限！

19.2　处理溢出

代码清单 19-2 展示了另一个版本，其中我们对字符进行计数，并且会忽略多余的字符。作为一个额外的调整，我们只允许使用小写字母字符 a 到 z。

代码清单 19-2：console2.asm

```
; console2.asm
section .data
        msg1    db      "Hello, World!",10,0
        msg2    db      "Your turn (only a-z): ",0
        msg3    db      "You answered: ",0
        inputlen equ 10         ;inputbuffer 的长度
        NL      db      0xa
section .bss
        input   resb    inputlen+1      ;为终止 0 提供空间
section .text
        global main
main:
push rbp
mov rbp,rsp
        mov     rdi, msg1       ; 打印第一个字符串
        call    prints
        mov     rdi, msg2       ; 打印第二个字符串，没有换行
        call    prints
        mov     rdi, input      ; inputbuffer 的地址
        mov     rsi, inputlen   ; inputbuffer 的长度
        call    reads           ; 等待输入
        mov     rdi, msg3       ; 打印第三个字符串并添加输入字符串
        call    prints
        mov     rdi, input      ; 打印 inputbuffer
        call    prints
        mov     rdi,NL          ; 打印换行
        call    prints
leave
ret
;----------------------------------------------------------------
prints:
push rbp
mov rbp, rsp
push r12                ; 被调用者保留

; 字符计数
        xor     rdx, rdx        ; 长度存放在 rdx 中
```

第 19 章 控制台 I/O

```
            mov     r12, rdi
    .lengthloop:
            cmp     byte [r12], 0
            je      .lengthfound
            inc     rdx
            inc     r12
            jmp     .lengthloop
    .lengthfound:               ; 打印字符串，长度在 rdx 中
            cmp     rdx, 0      ; 没有字符串（长度为 0）
            je      .done
            mov     rsi,rdi     ; rdi 包含字符串的长度
            mov     rax, 1      ; 1 表示写入
            mov     rdi, 1      ; 1 表示标准输出
            syscall
    .done:
    pop r12
    leave
    ret
    ;-----------------------------------------------------
    reads:
    section .data
    section .bss
            .inputchar  resb 1
    section .text
    push rbp
    mov rbp, rsp
            push    r12             ; 被调用者保留
            push    r13             ; 被调用者保留
            push    r14             ; 被调用者保留
            mov     r12, rdi        ; inputbuffer 的地址
            mov     r13, rsi        ; 最大长度存放在 r13 中
            xor     r14, r14        ; 字符计数器
    .readc:
            mov     rax, 0          ; 读
            mov     rdi, 1          ; 标准输入
            lea     rsi, [.inputchar]   ; 输入的地址
            mov     rdx, 1          ; 读取的第几个字符
            syscall
            mov     al, [.inputchar]    ; 字符是换行符吗？
```

```
            cmp   al, byte[NL]
            je    .done           ; 换行结束
            cmp   al, 97          ; 比 a 还小吗?
            jl    .readc          ; 忽略
            cmp   al, 122         ; 比 z 还大吗?
            jg    .readc          ; 忽略
            inc   r14             ; 增加计数器
            cmp   r14, r13
            ja    .readc          ; 达到缓冲区的最大限制,忽略
            mov   byte [r12], al  ; 安全缓冲区中的字符
            inc   r12             ; 指向缓冲区中的下一个字符
            jmp   .readc
    .done:
            inc   r12
            mov byte [r12],0      ; 为 inputbuffer 增加终止 0
            pop r14               ; 被调用者保留
            pop r13               ; 被调用者保留
            pop r12               ; 被调用者保留
    leave
    ret
```

我们修改了 prints 函数,以便它首先计算要显示的字符数。也就是说,它一直计数,直至找到一个 0 字节为止。确定长度后,prints 将使用 syscall 显示字符串。

reads 函数等待输入一个字符,并检查它是否为新行。如果是新行,则停止从键盘读取字符。寄存器 r14 保存输入字符的计数。该函数检查字符数是否大于 inputlen;如果不是,则将字符添加到缓冲区 input 中。如果大于 inputlen,则忽略该字符,但继续从键盘读取。我们要求字符的 ASCII 码在 97 和 122 之间。这将确保仅接收小写字母字符。请注意,我们保存并恢复了被调用方保存的寄存器;我们在两个函数(prints 和 reads)中都使用了 r12。这种情况下,不保存被调用方保存的寄存器将不是问题,但你可以想象,如果第一个函数调用第二个函数而第二个又调用第三个函数,则可能出现问题。

图 19-3 展示了输出。

```
jo@ubuntu18:~/Desktop/Book/24 console 2$ make
nasm -f elf64 -g -F dwarf console2.asm -l console2.lst
gcc -o console2 console2.o -no-pie
jo@ubuntu18:~/Desktop/Book/24 console 2$ ./console2
Hello, World!
Your turn (only a-z): 123a{bcde}fghijklmnop
You answered: abcdefghij
jo@ubuntu18:~/Desktop/Book/24 console 2$
```

图 19-3　console2.asm 的输出

使用 SASM 调试控制台输入非常复杂，因为我们通过 syscall 提供输入。SASM 为 I/O 提供了自己的功能，但我们不想使用它，因为我们希望在不隐藏细节的情况下展示汇编语言和机器语言的工作方式。如果你陷于 SASM 调试中，请使用我们的老朋友 GDB。

19.3　小结

本章内容：
- 使用 syscall 进行键盘输入
- 验证键盘输入
- 调试键盘输入，这可能十分复杂

第 20 章

文件 I/O

在软件开发中，文件操作可能很复杂。不同操作系统具有不同的文件管理方法，每种方法都有一系列不同的选项。在本章中，我们将讨论 Linux 系统的文件 I/O。你将在第 43 章中看到，Windows 中的文件 I/O 是完全不同的。

在 Linux 中，文件管理非常复杂，涉及创建和打开文件以进行只读或读/写，写入新文件或附加到文件以及删除文件。更不用说"用户""组"和"其他"的安全设置了。如有必要，请提高你对 Linux 文件系统的管理技能，然后擦掉 Linux 系统管理手册上的灰尘来刷新一下你的记忆。我们在代码中仅指定了当前"用户"的标志，但是也可以添加"组"和"其他"的标志。如果你不知道我们在说什么，那么现在是该学习一些有关基本 Linux 文件管理的知识了。

20.1 使用 syscall

使用 syscall 创建、打开、关闭文件等。在本章中，我们将使用很多 syscall，因此我们将简化一些事情。在代码开头，我们将定义比 syscall 调用更易于引用的常量。你可在以下代码中识别 syscall 常量，因为它们以 NR_ 开头。使用这些常量可以使代码更具可读性。你可在系统的以下文件中找到 syscall 符号名称的列表：

```
/usr/include/asm/unistd_64.h
```

我们将在程序中使用相同的名称。注意，还有一个名为 unistd_32h 的文件，用于对 32 位的兼容。

我们还为创建标志、状态标志和访问模式标志创建了符号常量。这些标志指示是否要创建或追加、只读、只写文件等。你可在系统的文件中找到这些标志的列表和描述。

```
/usr/include/asm-generic/fcntl.h
```

这些标志以八进制表示(例如，O_CREAT=00000100)。以 0x 开头的值是十六进制值，以 0 开头但没有 x 的值是八进制值。为了提高可读性，你可将字符 q 附加到八进制数。

创建文件时，必须指定文件权限。请记住，在 Linux 中，你拥有用户、组和其他用户的读、写和执行权限。使用以下 CLI 命令，你可以查看概述并发现许多细微之处：

```
man 2 open
```

文件权限也以八进制表示法给出，Linux 系统管理员对此很熟悉。为了保持一致，我们将借用这些文件中使用的符号名称。

示例程序很长，但是我们将逐步分析它，可以使用条件汇编来完成。这使你有机会逐段分析程序。

20.2 文件处理

在程序中执行以下操作：
(1) 创建一个文件，然后在文件中写入数据。
(2) 覆盖文件内容的一部分。
(3) 将数据追加到文件。
(4) 将数据写入文件中的特定位置。
(5) 从文件中读取数据。
(6) 从文件中的特定位置读取数据。
(7) 删除文件。

代码清单 20-1 展示了代码。

代码清单 20-1：file.asm

```
; file.asm
section .data
; 用于条件汇编的表达式
        CREATE      equ 1
        OVERWRITE   equ 1
        APPEND      equ 1
        O_WRITE     equ 1
        READ        equ 1
```

```
        O_READ      equ 1
        DELETE      equ 1
; 系统调用符号
        NR_read     equ 0
        NR_write    equ 1
        NR_open     equ 2
        NR_close    equ 3
        NR_lseek    equ 8
        NR_create   equ 85
        NR_unlink   equ 87
; 创建和状态标志
        O_CREAT     equ 00000100q
        O_APPEND    equ 00002000q
; 访问模式
        O_RDONLY    equ 000000q
        O_WRONLY    equ 000001q
        O_RDWR      equ 000002q
; 创建模式 (权限)
        S_IRUSR     equ 00400q      ;用户读取权限
        S_IWUSR     equ 00200q      ;用户写入权限

        NL      equ 0xa
        bufferlen   equ 64
        fileName    db "testfile.txt",0
        FD      dq 0        ; 文件描述符

        text1   db "1. Hello...to everyone!",NL,0
        len1    dq $-text1-1        ;移除 0
        text2   db "2. Here I am!",NL,0
        len2    dq $-text2-1        ;移除 0
        text3   db "3. Alife and kicking!",NL,0
        len3    dq $-text3-1        ;移除 0
        text4   db "Adios !!!",NL,0
    len4        dq $-text4-1

    error_Create    db "error creating file",NL,0
    error_Close     db "error closing file",NL,0
    error_Write     db "error writing to file",NL,0
```

```nasm
        error_Open     db "error opening file",NL,0
        error_Append   db "error appending to file",NL,0
        error_Delete   db "error deleting file",NL,0
        error_Read     db "error reading file",NL,0
        error_Print    db "error printing string",NL,0
        error_Position db "error positioning in file",NL,0

        success_Create   db "File created and opened",NL,0
        success_Close    db "File closed",NL,NL,0
        success_Write    db "Written to file",NL,0
        success_Open     db "File opened for R/W",NL,0
        success_Append   db "File opened for appending",NL,0
        success_Delete   db "File deleted",NL,0
        success_Read     db "Reading file",NL,0
        success_Position db "Positioned in file",NL,0

section .bss
        buffer  resb bufferlen
section .text
        global main
main:
        push rbp
        mov rbp,rsp
%IF CREATE
;创建并打开文件,然后关闭 ---------------------
; 创建并打开文件
        mov  rdi, fileName
        call createFile
        mov  qword [FD], rax    ; 保存描述符

; 写入文件#1
        mov rdi, qword [FD]
        mov rsi, text1
        mov rdx, qword [len1]
        call writeFile

; 关闭文件
        mov rdi, qword [FD]
        call closeFile
%ENDIF
%IF OVERWRITE
```

```
;打开并覆盖文件，然后关闭 -----------------
; 打开文件
        mov rdi, fileName
        call openFile
        mov qword [FD], rax     ; 保存文件描述符

; 写入文件#2，覆盖！
        mov rdi, qword [FD]
        mov rsi, text2
        mov rdx, qword [len2]
        call writeFile

; 关闭文件
        mov rdi, qword [FD]
        call closeFile
%ENDIF
%IF APPEND
;打开并附加到文件，然后关闭 ----------------
; 打开要追加的文件
        mov rdi, fileName
        call appendFile
        mov qword [FD], rax     ; 保存文件描述符

; 写入 #3，追加！
        mov rdi, qword [FD]
        mov rsi, text3
        mov rdx, qword [len3]
        call writeFile

; 关闭文件
        mov rdi, qword [FD]
        call closeFile
%ENDIF
%IF O_WRITE
;打开并覆盖文件中的偏移量，然后关闭 ----
; 打开文件进行写入
        mov rdi, fileName
        call openFile
        mov qword [FD], rax     ; 保存文件描述符

; 从偏移位置定位文件
```

```
        mov rdi, qword[FD]
        mov rsi, qword[len2]        ;这个位置的偏移量
        mov rdx, 0
        call positionFile

; 从偏移位置写入文件
        mov rdi, qword[FD]
        mov rsi, text4
        mov rdx, qword [len4]
        call writeFile

; 关闭文件
        mov rdi, qword [FD]
        call closeFile
%ENDIF
%IF READ
; 打开并读取文件,然后关闭 -----------------
; 打开文件进行读取
        mov rdi, fileName
        call openFile
        mov qword [FD], rax         ; 保存文件描述符

; 从文件读取
        mov rdi, qword [FD]
        mov rsi, buffer
        mov rdx, bufferlen
        call readFile
        mov rdi,rax
        call printString

; 关闭文件
        mov rdi, qword [FD]
        call closeFile
%ENDIF
%IF O_READ
;从文件的偏移处打开并读取,然后关闭 -----
; 打开文件进行读取
        mov rdi, fileName
        call openFile
        mov qword [FD], rax         ; 保存文件描述符
```

```nasm
; 使用偏移量定位文件
        mov rdi, qword[FD]
        mov rsi, qword[len2]     ;跳过第一行
        mov rdx, 0
        call positionFile

; 从文件读取
        mov rdi, qword [FD]
        mov rsi, buffer
        mov rdx, 10              ;要读取的字符数
        call readFile
        mov rdi,rax
        call printString

; 关闭文件
        mov rdi, qword [FD]
        call closeFile
%ENDIF
%IF DELETE
;删除一个文件 --------------------------------
; 删除文件，取消注释要使用的接下来的一些行
        mov rdi, fileName
        call deleteFile
%ENDIF

leave
ret

; 文件操作函数--------------------
;---------------------------------------------
global readFile
readFile:
        mov   rax, NR_read
        syscall                  ; rax 包含读取的字符数
        cmp   rax, 0
        jl    readerror
        mov   byte [rsi+rax],0   ; 添加一个终止 0
        mov   rax, rsi
        mov   rdi, success_Read
        push  rax                ; 调用者保留
        call  printString
```

```
        pop     rax         ; 调用者保留
        ret
readerror:
        mov     rdi, error_Read
        call    printString
        ret
;-------------------------------------------------
global deleteFile
deleteFile:
        mov rax, NR_unlink
        syscall
        cmp rax, 0
        jl  deleteerror
        mov rdi, success_Delete
        call printString
        ret
deleteerror:
        mov rdi, error_Delete
        call printString
        ret
;-------------------------------------------------
global appendFile
appendFile:
        mov     rax, NR_open
        mov     rsi, O_RDWR|O_APPEND
        syscall
        cmp     rax, 0
        jl      appenderror
        mov     rdi, success_Append
        push    rax         ; 调用者保留
        call    printString
        pop     rax         ; 调用者保留
        ret
appenderror:
        mov rdi, error_Append
        call printString
        ret
;-------------------------------------------------
global openFile
```

```
openFile:
    mov rax, NR_open
    mov rsi, O_RDWR
    syscall
    cmp rax, 0
    jl  openerror
    mov rdi, success_Open
    push rax         ; 调用者保留
    call printString
    pop rax          ; 调用者保留
    ret
openerror:
    mov rdi, error_Open
    call printString
    ret
;--------------------------------------------
global writeFile
writeFile:
    mov  rax, NR_write
    syscall
    cmp  rax, 0
    jl   writeerror
    mov  rdi, success_Write
    call printString
    ret
writeerror:
    mov rdi, error_Write
    call printString
    ret
;--------------------------------------------
global positionFile
positionFile:
    mov  rax, NR_lseek
    syscall
    cmp  rax, 0
    jl   positionerror
    mov  rdi, success_Position
    call printString
    ret
```

```
positionerror:
        mov   rdi, error_Position
        call  printString
        ret
;------------------------------------------------
global closeFile
closeFile:
        mov   rax, NR_close
        syscall
        cmp   rax, 0
        jl    closeerror
        mov   rdi, success_Close
        call  printString
        ret
closeerror:
        mov   rdi, error_Close
        call  printString
        ret
;------------------------------------------------
global createFile
createFile:
        mov   rax, NR_create
        mov   rsi, S_IRUSR |S_IWUSR
        syscall
        cmp   rax, 0            ; 文件描述符保存在 rax 中
        jl    createerror
        mov   rdi, success_Create
        push  rax               ; 调用者保留
        call  printString
        pop   rax               ; 调用者保留
        ret
createerror:
        mov rdi, error_Create
        call  printString
        ret
; 打印反馈
;------------------------------------------------
global printString
printString:
```

```
        ; 字符计数
              mov   r12, rdi
              mov   rdx, 0
       strLoop:
              cmp   byte [r12], 0
              je    strDone
              inc   rdx            ;长度保存在 rdx 中
              inc   r12
              jmp   strLoop
       strDone:
              cmp   rdx, 0         ; 没有字符串(长度为 0)
              je    prtDone
              mov   rsi,rdi
              mov   rax, 1
              mov   rdi, 1
              syscall
       prtDone:
              ret
```

20.3 条件汇编

因为这是一个相当长的程序，为了便于分析，我们使用了条件汇编。我们创建了不同的常量，如 CREATE、WRITE、APPEND 等。如果你将这样一个变量设置为 1，则某些代码(由%IF 'variable'和%ENDIF 括起来)将被汇编。如果该变量设置为 0，则汇编程序将忽略该代码。%IF 和%ENDIF 部分称为汇编预处理器指令。从变量 CREATE equ 1 开始，然后将其他变量设置为 0，再汇编、运行和分析程序。之后继续创建 CREATE equ 1 和 OVERWRITE equ 1，并在第二次编译中将其他变量设置为 0，以此类推。

NASM 为我们提供了大量的预处理指令。这里使用条件汇编。如前所述，为了定义宏，我们还使用了预处理指令。在 NASM 手册的第 4 章中，你将找到预处理指令的完整描述。

20.4 文件操作指令

下面首先创建一个文件。将文件名移到 rdi，然后调用 createFile。在 createFile 中，将符号变量 NR_create 放入 rax，并在 rsi 中指定用于创建文件的标志。这种情况下，用户将获得读写权限，然后执行 syscall。

由于某种原因无法创建文件时，createFile 在 rax 中返回一个负值，这种情况下，我们希望显示错误消息。如果需要更多详细信息，rax 中的负值将表示发生了哪种错误。如果创建了文件，则该函数在 rax 中返回一个文件描述符。在调用程序中，我们将文件描述符保存到变量 FD 中，以进一步执行文件操作。你可以看到，在调用 printString 函数之前，我们必须小心保留 rax 的内容。调用 printString 会破坏 rax 的内容，因此我们在调用之前将 rax 压入堆栈。根据调用约定，rax 是一个调用方保存的寄存器。

代码的下一步是将一些文本写入文件，然后关闭文件。请注意，当你创建文件时，将创建一个新文件。如果存在同名文件，则将其删除。

使用 CREATE equ 1 编译并运行程序；其他条件汇编变量等于 0。然后转到命令行提示符，并验证是否创建了 testfile.txt 文件，以及其中包含的信息。如果要查看十六进制文件的内容(有时很有用)，请在命令行提示符下使用 xxd testfile.txt。

逐步将条件汇编变量设置为 1(一次一个)，然后检查 testfile.txt 发生了什么。

注意，这种情况下，我们创建并使用了没有函数序言和尾声的函数。图 20-1 展示了输出，所有条件汇编变量都设置为 1。

```
jo@UbuntuDesktop:~/Desktop/linux64/gcc/25 file$ make
nasm -f elf64 -g -F dwarf file.asm -l file.lst
gcc -o file file.o -no-pie
jo@UbuntuDesktop:~/Desktop/linux64/gcc/25 file$ ./file
File created and opened
Written to file
File closed

File opened for reading/(over)writing/updating
Written to file
File closed

File opened for appending
Written to file
File closed

File opened for reading/(over)writing/updating
Positioned in file
Written to file
File closed

File opened for reading/(over)writing/updating
Reading file
2. Here I am!
Adios !!!
3. Alife and kicking!
File closed

File opened for reading/(over)writing/updating
Positioned in file
Reading file
Adios !!!
File closed

File deleted
jo@UbuntuDesktop:~/Desktop/linux64/gcc/25 file$
```

图 20-1　file.asm 的输出

20.5　小结

本章内容：
- 创建、打开、关闭、删除文件
- 写入文件、追加文件以及从一个特定位置写入文件
- 读取文件
- 文件操作的不同参数

第 21 章

命令行

有时你希望在命令行中启动带有参数的程序。这在开发自己的命令行工具时很有用。系统管理员一直都在使用命令行工具。因为一般来说，对于技术功底深厚的用户，命令行工具效率更高。

21.1 访问命令行参数

在代码清单 21-1 的示例程序中，我们展示了如何在汇编程序中访问命令行参数。为了尽量保持简单，我们只是找到参数并打印出来。

代码清单 21-1：cmdline.asm

```
;cmdline.asm
extern printf
section .data
      msg db "The command and arguments: ",10,0
      fmt db "%s",10,0
section .bss
section .text
      global main
main:
push rbp
mov rbp,rsp
      mov r12, rdi       ; 参数数量
      mov r13, rsi       ; 参数数组的地址
; 打印标题
    mov   rdi, msg
```

```
        call    printf
        mov     r14, 0
    ; 打印命令和参数
    .ploop:                  ; 循环遍历整个数组并打印
        mov     rdi, fmt
        mov     rsi, qword [r13+r14*8]
        call    printf
        inc     r14
        cmp     r14, r12      ; 参数数量达到了吗?
        jl      .ploop
    leave
    ret
```

执行该程序时，参数数量(包括程序名称本身)存储在 rdi 中。寄存器 rsi 包含内存中数组的地址，其中包含命令行参数的地址，第一个参数是程序本身。使用 rdi 和 rsi 符合调用约定。请记住，我们在 Linux 上使用 System V AMD64 ABI 调用约定进行工作；在其他平台(例如 Windows)上，使用其他调用约定。我们复制此信息是因为 rdi 和 rsi 稍后将用于 printf。

代码循环遍历参数数组，直到达到参数总数为止。在.ploop 循环中，r13 指向参数数组。寄存器 r14 用作自变量计数器。在每个循环中，将计算下一个参数的地址并将其存储在 rsi 中。qword [r13 + r14 * 8]中的 8 表示所指向地址的长度：8 字节×8 位= 64 位地址。在每个循环中，将寄存器 r14(包含参数的数量)与 r12 进行比较。

图 21-1 展示了带有一些随机参数的输出。

```
jo@UbuntuDesktop:~/Desktop/linux64/gcc/26 cmdline$ make
nasm -f elf64 -g -F dwarf cmdline.asm -l cmdline.lst
gcc -o cmdline cmdline.o -no-pie
jo@UbuntuDesktop:~/Desktop/linux64/gcc/26 cmdline$ ./cmdline arg1 arg2 abc 5
The command and arguments:
./cmdline
arg1
arg2
abc
5
jo@UbuntuDesktop:~/Desktop/linux64/gcc/26 cmdline$
```

图 21-1　cmdline.asm 的输出

21.2　调试命令行

目前，SASM 不能用于调试带有命令行参数的程序；你必须使用 GDB。以下是执行此操作的一种方法。

```
gdb --args ./cmdline arg1 arg2 abc 5
break main
run
info registers rdi rsi rsp
```

你可以使用前面的指令验证 rdi 是否包含参数的数量(包括命令本身)，以及 rsi 是否指向高内存中的地址，甚至高于堆栈，如第 8 章所述。图 21-2 展示了 GDB 的输出。

```
(gdb) break main
Breakpoint 1 at 0x4004a0: file cmdline.asm, line 9.
(gdb) run
Starting program: /home/jo/Desktop/linux64/gcc/26 cmdline/cmdline arg1 arg2 abc 5

Breakpoint 1, main () at cmdline.asm:9
9         push rbp
(gdb) info registers rdi rsi rsp
rdi            0x5      5
rsi            0x7fffffffde58   140737488346712
rsp            0x7fffffffdd78   0x7fffffffdd78
(gdb)
```

图 21-2　GDB cmdline 输出

在图 21-2 中，带有参数地址的数组从 0x7fffffde58 开始。下面更深入地了解实际的参数。第一个参数的地址如下所示：

```
x/1xg 0x7fffffffde58
```

这里我们要求在地址 0x7fffffde58 处输入一个十六进制的 giant 字(8 字节)。图 21-3 展示了答案。

```
(gdb) x/1xg 0x7fffffffde58
0x7fffffffde58: 0x00007fffffffe204
(gdb)
```

图 21-3　GDB 第一个参数的地址

下面找出该地址的内容(图 21-4)。

```
x/s 0x7fffffffe204
```

```
(gdb) x/s 0x7fffffffe204
0x7fffffffe204: "/home/jo/Desktop/linux64/gcc/26 cmdline/cmdline"
(gdb)
```

图 21-4　GDB(第一个参数)

这确实是我们的第一个参数，即命令本身。要查找第二个参数，请使用 8 个字节将 0x7fffffffde58 扩充为 0x7fffffffde60，找到第二个参数的地址，以此类推。结果如图 21-5 所示。

```
(gdb) x/1xg 0x7fffffffde60
0x7fffffffde60: 0x00007fffffffe234
(gdb) x/s 0x7fffffffe234
0x7fffffffe234: "arg1"
(gdb)
```

图 21-5　GDB(第二个参数)

这是调试和验证命令行参数的方法。

21.3　小结

本章内容：
- 如何访问命令行参数
- 如何对命令行参数应用寄存器
- 如何调试命令行参数

第 22 章

从 C 到汇编

在前面的章节中，为了方便起见，我们经常使用 C 函数，例如标准 printf 函数或我们开发的 printb 版本。在本章中，我们将展示如何在 C 语言中使用汇编函数。调用约定的价值将立即显现出来。在本章中，我们将使用 System V AMD64 ABI 调用约定，因为我们正在 Linux 操作系统上工作。Windows 有不同的调用约定。如果你已经完成了前面的章节和示例代码，那么本章将很容易理解。

22.1 编写 C 源文件

在前面的章节中，你应该熟悉了大多数汇编代码。只有 C 程序是新的。我们计算矩形和圆形的面积和周长。然后取一个字符串并反转它，最后取一个数组的元素之和，将数组的元素加倍，然后取加倍数组的元素之和。下面看看不同的源文件。

下面从 C 源文件开始，参见代码清单 22-1。

代码清单 22-1：fromc.c.asm

```
// fromc.c

#include <stdio.h>
#include <string.h>

extern int rarea(int, int);
extern int rcircum(int, int);
extern double carea( double);
extern double ccircum( double);
extern void sreverse(char *, int );
```

```c
extern void adouble(double [], int );
extern double asum(double [], int );
int main()
{
    char rstring[64];
    int side1, side2, r_area, r_circum;
    double radius, c_area, c_circum;
    double darray[] = {70.0, 83.2, 91.5, 72.1, 55.5};
    long int len;
    double sum;

    // 使用 int 参数调用汇编函数
    printf("Compute area and circumference of a rectangle\n");
    printf("Enter the length of one side : \n");
    scanf("%d", &side1 );
    printf("Enter the length of the other side : \n");
    scanf("%d", &side2 );

    r_area = rarea(side1, side2);
    r_circum = rcircum(side1, side2);

    printf("The area of the rectangle = %d\n", r_area);
    printf("The circumference of the rectangle = %d\n\n", r_circum);

    // 使用双精度浮点数参数调用汇编函数
    printf("Compute area and circumference of a circle\n");
    printf("Enter the radius : \n");
    scanf("%lf", &radius);

    c_area = carea(radius);
    c_circum = ccircum(radius);
    printf("The area of the circle = %lf\n", c_area);
    printf("The circumference of the circle = %lf\n\n", c_circum);

    // 使用字符串参数调用汇编函数
    printf("Reverse a string\n");
    printf("Enter the string : \n");
    scanf("%s", rstring);
    printf("The string is = %s\n", rstring);
```

```c
        sreverse(rstring,strlen(rstring));
        printf("The reversed string is = %s\n\n", rstring);

// 使用数组参数调用汇编函数
        printf("Some array manipulations\n");
        len = sizeof (darray) / sizeof (double);

        printf("The array has %lu elements\n",len);
        printf("The elements of the array are: \n");
        for (int i=0;i<len;i++){
            printf("Element %d = %lf\n",i, darray[i]);
        }

        sum = asum(darray,len);
        printf("The sum of the elements of this array = %lf\n", sum);

        adouble(darray,len);
        printf("The elements of the doubled array are: \n");
        for (int i=0;i<len;i++){
            printf("Element %d = %lf\n",i, darray[i]);
        }

        sum = asum(darray,len);
        printf("The sum of the elements of this doubled array = %lf\n", sum);
        return 0;
}
```

22.2 编写汇编代码

我们从汇编函数的函数声明开始。这些是外部函数，我们声明返回值和参数的数据类型。

该程序将提示用户要使用的大多数数据，但数组除外，在数组中为方便起见，我们提供了一些值。

代码清单 22-2 至代码清单 22-7 展示了汇编函数。

代码清单 22-2：rect.asm

```
;rect.asm
section .data
```

```
section .bss
section .text

global rarea
rarea:
        section .text
                push    rbp
                mov     rbp, rsp
                mov     rax, rdi
                imul    rsi
                leave
                ret
global rcircum
rcircum:
        section .text
                push    rbp
                mov     rbp, rsp
                mov     rax, rdi
                add     rax, rsi
                imul    rax, 2
                leave
                ret
```

代码清单 22-3：circle.asm

```
;circle.asm
section .data
     pi dq 3.141592654

section .bss
section .text
global carea
carea:
        section .text
                push    rbp
                mov     rbp, rsp
                movsd   xmm1, qword [pi]
                mulsd   xmm0,xmm0        ; 半径存放在 xmm0 中
                mulsd   xmm0, xmm1
```

```
                leave
                ret
global ccircum
ccircum:
        section .text
            push    rbp
            mov     rbp, rsp
            movsd   xmm1, qword [pi]
            addsd   xmm0,xmm0          ; 半径存放在 xmm0 中
            mulsd   xmm0, xmm1
            leave
            ret
```

代码清单 22-4：sreverse.asm

```
;sreverse.asm
section .data
section .bss
section .text

global sreverse
sreverse:
push rbp
mov rbp, rsp
pushing:

        mov rcx, rsi
        mov rbx, rdi
        mov r12, 0
        pushLoop:
            mov rax, qword [rbx+r12]
            push rax
            inc r12
            loop pushLoop
popping:
        mov rcx, rsi
        mov rbx, rdi
        mov r12, 0
        popLoop:
```

```
            pop rax
            mov byte [rbx+r12], al
            inc r12
            loop popLoop
    mov rax, rdi
    leave
    ret
```

代码清单 22-5：asum.asm

```
; asum.asm
section .data
section .bss
section .text

global asum
asum:
        section .text
;calculate the sum
            mov rcx, rsi      ;数组长度
            mov rbx, rdi      ;数组地址
            mov r12, 0
            movsd   xmm0, qword [rbx+r12*8]
            dec rcx           ; 至少循环一次，第一个元素已经在 xmm0 中
    sloop:
            inc r12
            addsd xmm0, qword [rbx+r12*8]
            loop sloop
    ret            ; 在 xmm0 中返回总和
```

代码清单 22-6：adouble.asm

```
; adouble.asm
section .data
section .bss
section .text
global adouble
adouble:
        section .text
;double the elements
```

```
                mov rcx, rsi        ; 数组长度
                mov rbx, rdi        ; 数组地址
                mov r12, 0
         aloop:
                movsd xmm0, qword [rbx+r12*8]  ; 从数组中取出一个元素
                addsd xmm0,xmm0                ; 加倍
                movsd qword [rbx+r12*8], xmm0  ; 把它移到数组
                inc r12
                loop aloop
ret
```

代码清单 22-7：makefile

```
fromc: fromc.c rect.o circle.o sreverse.o adouble.o asum.o
       gcc -o fromc fromc.c rect.o circle.o sreverse.o \
       adouble.o asum.o -no-pie
rect.o: rect.asm
       nasm -f elf64 -g -F dwarf rect.asm -l rect.lst
circle.o: circle.asm
       nasm -f elf64 -g -F dwarf circle.asm -l circle.lst
sreverse.o: sreverse.asm
       nasm -f elf64 -g -F dwarf sreverse.asm -l sreverse.lst
adouble.o: adouble.asm
       nasm -f elf64 -g -F dwarf adouble.asm -l adouble.lst
asum.o: asum.asm
       nasm -f elf64 -g -F dwarf asum.asm -l asum.lst
```

在汇编代码中没有什么特别之处。请注意从调用 C 程序接收到的变量的数据类型。汇编函数从调用程序中获取参数，然后根据调用约定将它们存储在寄存器中。结果以 rax(整数值)或 xmm0(浮点值)的形式返回给调用方。现在，你可以开发自己的函数库以用于汇编程序或 C 中；并且由于调用约定，你不必担心如何传递参数。请注意使用正确的数据类型。

请注意我们如何在 makefile 中使用反斜杠(\)来分隔长行，并使用制表符来对齐指令。

图 22-1 展示了输出。

```
jo@ubuntu18:~/Desktop/Book/27 fromc$ ./fromc
Compute area and circumference of a rectangle
Enter the length of one side :
2
Enter the length of the other side :
3
The area of the rectangle = 6
The circumference of the rectangle = 10

Compute area and circumference of a circle
Enter the radius :
10
The area of the circle = 314.159265
The circumference of the circle = 62.831853

Reverse a string
Enter the string :
abcde
The string is = abcde
The reversed string is = edcba

Double the elements of an array
The array has 5 elements
The elements of the array are:
Element 0 = 70.000000
Element 1 = 83.200000
Element 2 = 91.500000
Element 3 = 72.100000
Element 4 = 55.500000
The sum of the elements of this array = 372.300000
The elements of the doubled array are:
Element 0 = 140.000000
Element 1 = 166.400000
Element 2 = 183.000000
Element 3 = 144.200000
Element 4 = 111.000000
The sum of this doubled array = 744.600000
jo@ubuntu18:~/Desktop/Book/27 fromc$
```

图 22-1　fromc.c 的输出

22.3　小结

本章内容：

- 从高级语言中调用汇编函数，在本例中是从 C 中调用汇编函数
- 调用约定的价值

第 23 章

内联汇编

我们将在本章使用 C 编程语言来解释内联汇编。可以在 C 程序中编写汇编指令。在大多数情况下，这是不可取的，因为当今的 C 编译器设计得非常好，以至于你需要成为一个非常熟练的汇编程序员才能提高 C 代码的性能。实际上，使用内联汇编会使 C 或 C++编译器更难以优化包含内联汇编的代码。

而且，C 编译器不会对汇编指令执行任何错误检查；你必须自己找出所有错误。此外，访问 C 程序正在使用的内存和寄存器可能会给自身带来风险。然而，互联网上的许多文章使用带内联汇编的 C 来解释底层功能，因此了解如何读取代码可能会很有用。

内联汇编有两种：基本和扩展。

23.1 基本内联汇编

下面从一个基本内联汇编示例开始。参见代码清单 23-1 和代码清单 23-2。

代码清单 23-1：inline1.c

```c
// inline1.c
#include <stdio.h>

int x=11, y=12, sum, prod;
int subtract(void);
void multiply(void);

int main(void)
{
```

```c
        printf("The numbers are %d and %d\n",x,y);
        __asm__(
            ".intel_syntax noprefix;"
            "mov rax,x;"
            "add rax,y;"
            "mov sum,rax"
        );
        printf("The sum is %d.\n",sum);
        printf("The difference is %d.\n",subtract());
        multiply();
        printf("The product is %d.\n",prod);
}

int subtract(void)
{
        __asm__(
            ".intel_syntax noprefix;"
            "mov rax,x;"
            "sub rax,y"        // 返回值进入 rax
        );
}
void multiply(void)
{
        __asm__(
            ".intel_syntax noprefix;"
            "mov rax,x;"
            "imul rax,y;"
            "mov prod,rax"     // 没有返回值，结果存放在 prod 中
        );
}
```

代码清单 23-2：makefile

```
# makefile inline1.c
inline1: inline1.c
    gcc -o inline1 inline1.c -masm=intel -no-pie
```

请注意 makefile 中的附加参数，即 -masm=intel。使用内联汇编时，此参数是必需的。

上一个示例显示了所谓的基本内联汇编程序。在主程序中，两个变量相加，然后调用一个函数减去两个变量，之后调用另一个函数乘以两个变量。如果要访问基本内联汇编程序中的变量，需要将它们声明为全局变量，即在任何函数外声明它们。如果变量不是全局变量，则 gcc 会找不到变量。但是全局变量易于出错，例如命名冲突。同样，当你在汇编代码中修改寄存器时，可能必须在调用内联汇编之前保存它们，并在退出内联汇编时将它恢复为原始值，否则可能导致程序崩溃。通过内联汇编修改的寄存器称为"破坏性寄存器"。

在包含__asm__(...)的汇编代码中，第一条语句表明我们要使用无前缀的 Intel 语法(请记住第 3 章中有关 Intel 语法和 AT&T 语法风格的讨论)。然后，我们像往常一样使用汇编指令，以 a 或\n 结尾。最后一个程序不必以;或\n 结尾。注意全局变量的使用。我们很幸运，因为破坏寄存器不会使程序崩溃。为了避免寄存器的混乱和使用全局变量，你需要使用扩展内联汇编，如下一节所述。

图 23-1 展示了输出。

```
jo@UbuntuDesktop:~/Desktop/linux64/gcc/28 inline 1$ make
gcc -o inline1 inline1.c -masm=intel
jo@UbuntuDesktop:~/Desktop/linux64/gcc/28 inline 1$ ./inline1
The numbers are 11 and 12
The sum is 23.
The difference is -1.
The product is 132.
jo@UbuntuDesktop:~/Desktop/linux64/gcc/28 inline 1$
```

图 23-1　inline1.c 的输出

23.2　扩展内联汇编

代码清单 23-3 和代码清单 23-4 展示了一个扩展内联汇编的示例。

代码清单 23-3：inline2.c

```
// inline2.c

#include <stdio.h>
    int a=12;     // 全局变量
    int b=13;
    int bsum;

int main(void)
{
printf("The global variables are %d and %d\n",a,b);
__asm__(
```

```c
        ".intel_syntax noprefix\n"
        "mov rax,a \n"
        "add rax,b \n"
        "mov bsum,rax \n"
        :::"rax"
        );

        printf("The extended inline sum of global variables is %d.\n\n", bsum);

    int x=14,y=16, esum, eproduct, edif;    // 局部变量

    printf("The local variables are %d and %d\n",x,y);
    __asm__(
        ".intel_syntax noprefix;"
        "mov rax,rdx;"
        "add rax,rcx;"
        :"=a"(esum)
        :"d"(x), "c"(y)
        );
        printf("The extended inline sum is %d.\n", esum);
    __asm__(
        ".intel_syntax noprefix;"
        "mov rbx,rdx;"
        "imul rbx,rcx;"
        "mov rax,rbx;"
        :"=a"(eproduct)
        :"d"(x), "c"(y)
        :"rbx"
        );
        printf("The extended inline product is %d.\n", eproduct);

    __asm__(
        ".intel_syntax noprefix;"
        "mov rax,rdx;"
        "sub rax,rcx;"
        :"=a"(edif)
        :"d"(x), "c"(y)
        );
        printf("The extended inline asm difference is %d.\n", edif);
```

}

代码清单 23-4：makefile

```
# makefile inline2.c
inline2: inline2.c
    gcc -o inline2 inline2.c -masm=intel -no-pie
```

汇编器指令看起来有所不同。具体来说，使用了一个模板，如下所示：

```
asm (
    assembler code
    : output operands              /* optional */
    : input operands               /* optional */
    : list of clobbered registers  /* optional */
);
```

在汇编代码后，将使用其他的和可选的信息。以上面代码中的嵌入式乘积为例(此处重复)：

```
__asm__(
    ".intel_syntax noprefix;"
    "mov rbx,rdx;"
    "imul rbx,rcx;"
    "mov rax,rbx;"
    :"=a"(eproduct)
    :"d"(x), "c"(y)
    :"rbx"
    );
    printf("The extended inline product is %d.\n", eproduct);
```

每个可选行均以冒号(:)开头，并且你必须遵守指令的顺序。a、d 和 c 称为寄存器约束，它们分别映射到寄存器 rax、rdx 和 rcx。这是寄存器约束映射到寄存器的方式：

```
a -> rax, eax, ax, al
b -> rbx, ebx, bx, bl
c -> rcx, ecx, cx, cl
d -> rdx, edx, dx, dl
S -> rsi, esi, si
D -> rdi, edi, di
```

```
r -> any register
```

第一个可选行中的:"=a"(eproduct)表示输出将为 rax，并且 rax 将引用变量 eproduct。寄存器 rdx 引用 x，而 rcx 引用 y，它们是输入变量。

最后请注意，rbx 在代码中被认为是被破坏的，将被恢复为原始值，因为它是在被破坏的寄存器列表中声明的。这种情况下，保持被破坏的状态不会使程序崩溃。它仅用于说明用法。在互联网上可以找到有关内联汇编的很多信息，但如上所述，仅在特定情况下才需要使用内联汇编。请记住，使用内联汇编将使 C 代码的可移植性降低。请参见图 23-2。

```
jo@UbuntuDesktop:~/Desktop/linux64/gcc/29  inline 2$ make
gcc -o inline2 inline2.c -masm=intel
jo@UbuntuDesktop:~/Desktop/linux64/gcc/29  inline 2$ ./inline2
The global variables are 12 and 13
The extended inline asm sum of global variables is 25.

The local variables are 14 and 16
The extended inline asm sum is 30.
The extended inline product is 224.
The extended inline asm difference is -2.
jo@UbuntuDesktop:~/Desktop/linux64/gcc/29  inline 2$
```

图 23-2   inline2.c 的输出

在后续章节中，将解释如何在 Windows 中使用汇编。Visual Studio 中的 x64 处理器不支持内联汇编。它仅在 x86 处理器上受支持。但 gcc 没有这个限制。

23.3 小结

本章内容：
- 基本内联汇编
- 扩展内联汇编

第 24 章

字符串

当想到字符串时,我们人类通常会假设字符串是一系列字符,这些字符构成了我们可以理解的单词或短语。但在汇编语言中,任何连续内存位置的列表或数组都被认为是一个字符串,不管它是不是人类可以理解的。汇编为我们提供了许多强大的指令,可有效地操作这些数据块。在我们的示例中将使用可读字符,但请记住,实际上汇编并不关心字符是否可读。我们将展示如何移动字符串、如何扫描它们以及如何比较它们。

尽管这些指令功能强大,但在后续章节中讨论 SIMD 指令时,我们将提供更好的功能。但是,下面从基本指令开始。

24.1　移动字符串

代码清单 24-1：move_strings.asm

```
; move_strings.asm
%macro prnt 2
    mov   rax, 1        ; 1 表示写入
    mov   rdi, 1        ; 1 表示标准输出
    mov   rsi, %1
    mov   rdx, %2
    syscall
      mov rax, 1
      mov rdi, 1
      mov rsi, NL
      mov rdx, 1
    syscall
```

```
%endmacro

section .data
        length          equ 95
        NL db 0xa
        string1 db "my_string of ASCII:"
        string2 db 10,"my_string of zeros:"
        string3 db 10,"my_string of ones:"
        string4 db 10,"again my_string of ASCII:"
        string5 db 10,"copy my_string to other_string:"
        string6 db 10,"reverse copy my_string to other_string:"
section .bss
        my_string resb length
        other_string resb length
section .text
        global main
main:
push rbp
mov rbp, rsp
;--------------------------------------------------
;用可打印的ascii字符填充字符串
            prnt string1,18
            mov rax,32
            mov rdi,my_string
            mov rcx, length
str_loop1:  mov byte[rdi], al      ; 简单的方法
            inc rdi
            inc al
            loop str_loop1
            prnt my_string,length
;--------------------------------------------------
;用ascii编码的0填充字符串
            prnt string2,20
            mov rax,48
            mov rdi,my_string
            mov rcx, length
str_loop2:  stosb              ; 不再需要inc rdi
            loop str_loop2
```

```
            prnt my_string,length
;------------------------------------------------
;用 ascii 编码的 1 填充字符串
            prnt string3,19
            mov rax, 49
            mov rdi,my_string
            mov rcx, length
            rep stosb         ; 不再需要 inc rdi 和循环
            prnt my_string,length
;------------------------------------------------
;用可打印的 ascii 字符再次填充字符串
            prnt string4,26
            mov rax,32
            mov rdi,my_string
            mov rcx, length
str_loop3:  mov byte[rdi], al   ; 简单的方法
            inc rdi
            inc al
            loop str_loop3
            prnt my_string,length
;------------------------------------------------
;将 my_string 复制到 other_string
            prnt string5,32
            mov rsi,my_string      ;rsi 是源
            mov rdi,other_string   ;rdi 是目的
            mov rcx, length
            rep movsb
            prnt other_string,length
;------------------------------------------------
;将 my_string 反向复制到 other_string
            prnt string6,40
            mov rax, 48       ;清除 other_string
            mov rdi,other_string
            mov rcx, length
            rep stosb
            lea rsi,[my_string+length-4]
            lea rdi,[other_string+length]
            mov rcx, 27       ;仅复制 27-1 个字符
            std               ;std 表示设置 DF，cld 表示清除 DF
```

```
          rep movsb
          prnt other_string,length
leave
ret
```

在这个程序中,我们使用宏(有关宏的更多细节,请参见第 18 章)进行打印,但也可使用 C 语言的 printf 函数,我们已经使用过很多次。

首先在 ASCII 表中创建一个包含 95 个可打印字符的字符串,第一个为 32(空格),最后为 126(波浪号或~)。我们首先打印一个标题,然后将第一个 ASCII 码放入 rax,让 rdi 指向内存中 my_string 的地址。然后,我们将字符串的长度放入 rcx 中以在循环中使用。在循环中,我们将一个 ASCII 代码从 al 复制到 my_string,获取下一个代码并将其写入 my_string 中的下一个内存地址,以此类推。最后,我们打印字符串。同样,这里没什么新内容。

在下一个区域中,我们将 my_string 的内容修改为全 0(ASCII 48)。为此,我们将字符串长度再次放入 rcx 中以构建循环。然后,我们使用 stosb 指令将 1(ASCII 49) 存储到 my_string。指令 stosb 仅需要 rdi 中的字符串的起始地址和要写入 rax 中的字符,并且 stosb 会在循环的每个重复步骤中移至下一个内存地址。我们不必再担心增加 rdi 了。

然后,我们更进一步,摆脱 rcx 循环。我们使用 rep stosb 指令将 stosb 重复多次。重复次数存储在 rcx 中。这是初始化内存的高效方法。

接下来,我们继续移动内存内容。严格来说,我们将复制内存块,而不是移动复制内容。首先,我们再次使用可读的 ASCII 代码初始化字符串。我们可以通过使用宏或函数来优化此代码,而不仅仅是重复代码。然后,我们开始复制字符串/内存块:从 my_string 到 other_string。源字符串的地址放入 rsi,目标字符串的地址放入 rdi。这很容易记住,因为 rsi 中的 s 代表源,而 rdi 中的 d 代表目的地。然后使用 rep movsb。当 rcx 变为 0 时,rep 复制停止。

在程序的最后一个区域,将反向移动内存内容。这个概念可能有点混乱,这里将详细介绍一下。在使用 movsb 时,会考虑 DF(方向标志)的内容。当 DF=0 时,rsi 和 rdi 增加 1,指向下一个更高的内存地址。当 DF=1 时,rsi 和 rdi 减小 1,指向下一个较低的内存地址。这意味着在 DF=1 的示例中,rsi 需要指向要复制的最高内存地址的地址,并从那里开始递减。此外,rdi 需要指向最高的目标地址并从那里减少。这样做的目的是在复制时"向后走",即在每个循环中减少 rsi 和 rdi。注意,rsi 和 rdi 都会减小;你不能使用 DF 增加一个寄存器并减少另一个寄存器(反转字符串)。在我们的示例中,我们不复制整个字符串,而仅复制小写字母,将它们放在目标位置的较高内存位置。指令 lea rsi,[my_string + length-4]在 rsi 中加载 my_string 的有效地址,并跳过不是字母的四个字符。可使用 std 将 DF 标志设置为 1,使用 cld 将 DF

标志设置为 0。然后，我们调用功能强大的 rep movsb。

当只有 26 个字符时，为什么要在 rcx 中放入 27 个呢？事实证明，在循环中执行其他任何操作之前，rep 会将 rcx 减少 1。你可使用 SASM 等调试器进行验证。注释掉对 prnt 宏的所有引用，以避免出现问题。你将看到，SASM 使你可以进入 rep 循环并验证内存和寄存器。当然，也可在英特尔手册中查找有关 rep 的信息。你可在"操作(Operation)"下找到如下内容：

```
IF AddressSize = 16
    THEN
            Use CX for CountReg;
            Implicit Source/Dest operand for memory use of SI/DI;
    ELSE IF AddressSize = 64
            THEN Use RCX for CountReg;
            Implicit Source/Dest operand for memory use of RSI/RDI;
    ELSE
            Use ECX for CountReg;
            Implicit Source/Dest operand for memory use of ESI/EDI;
FI;
WHILE CountReg =/ 0
        DO
                Service pending interrupts (if any);
                Execute associated string instruction;
                CountReg ← (CountReg - 1);
                IF CountReg = 0
                    THEN exit WHILE loop; FI;
                IF (Repeat prefix is REPZ or REPE) and (ZF = 0)
                or (Repeat prefix is REPNZ or REPNE) and (ZF = 1)
                    THEN exit WHILE loop; FI;
        OD;
```

CountReg ← (CountReg – 1);告诉我们，计数器将首先减少。研究指令的操作对于理解指令的行为可能很有用。最后，stosb 和 movsb 使用字节。还有 stosw、movsw、stosd 和 movsd 可以处理单字和双字，并且 rsi 和 rdi 相应地增加或减少，其中 1 表示字节，2 表示单字，4 表示双字。

图 24-1 展示了示例程序的输出。

```
jo@UbuntuDesktop:~/Desktop/linux64/gcc/30 strings 1$ make
nasm -f elf64 -g -F dwarf move_strings.asm -l move_strings.lst
gcc -o move_strings move_strings.o
jo@UbuntuDesktop:~/Desktop/linux64/gcc/30 strings 1$ ./move_strings
my_string of ASCII
 !"#$%&'()*+,-./0123456789:;<=>?@ABCDEFGHIJKLMNOPQRSTUVWXYZ[\]^_`abcdefghijklmnopqrstuvwxyz{|}~

my_string of zeros:
0000000000000000000000000000000000000000000000000000000000000000000000000000000000000000000000

my_string of ones:
1111111111111111111111111111111111111111111111111111111111111111111111111111111111111111111111

again my_string of ASCII:
 !"#$%&'()*+,-./0123456789:;<=>?@ABCDEFGHIJKLMNOPQRSTUVWXYZ[\]^_`abcdefghijklmnopqrstuvwxyz{|}~

copy my_string to other_string:
 !"#$%&'()*+,-./0123456789:;<=>?@ABCDEFGHIJKLMNOPQRSTUVWXYZ[\]^_`abcdefghijklmnopqrstuvwxyz{|}~

reverse copy my_string to other_string:
0000000000000000000000000000000000000000000000000000000abcdefghijklmnopqrstuvwxyz
jo@UbuntuDesktop:~/Desktop/linux64/gcc/30 strings 1$
```

图 24-1 move_strings.asm 的输出

24.2 比较和扫描字符串

移动和复制字符串很重要，但是分析字符串的能力也很重要。在代码清单 24-2 所示的示例代码中，我们使用 **cmpsb** 指令比较两个字符串，并使用 **scasb** 在字符串中查找特定字符。

代码清单 24-2：strings.asm

```
; strings.asm
extern printf
section .data
        string1     db "This is the 1st string.",10,0
        string2     db "This is the 2nd string.",10,0
        strlen2     equ $-string2-2
        string21    db "Comparing strings: The strings do not differ.",10,0
        string22    db "Comparing strings: The strings differ, "
                    db "starting at position: %d.",10,0
        string3     db "The quick brown fox jumps over the lazy dog.",0
        strlen3     equ $-string3-2
        string33    db "Now look at this string: %s",10,0
        string4     db "z",0
        string44    db "The character '%s' was found at position: %d.",10,0
        string45    db "The character '%s' was not found.",10,0
        string46    db "Scanning for the character '%s'.",10,0
section .bss
section .text
```

```
        global main
main:
        push rbp
        mov rbp,rsp
;打印两个字符串
        xor    rax,rax
        mov rdi, string1
        call printf
        mov rdi, string2
        call printf
;比较两个字符串----------------------------------------
        lea    rdi,[string1]
        lea    rsi,[string2]
        mov rdx, strlen2
        call compare1
        cmp rax,0
        jnz    not_equal1
;字符串相同,打印
        mov rdi, string21
        call printf
        jmp otherversion
;字符串不相同,打印
not_equal1:
        mov rdi, string22
        mov rsi, rax
        xor    rax,rax
        call printf
;比较两个字符串,其他版本-----------------------------------
otherversion:
        lea    rdi,[string1]
        lea    rsi,[string2]
        mov rdx, strlen2
        call compare2
        cmp rax,0
        jnz    not_equal2
;字符串相同,打印
        mov rdi, string21
        call printf
        jmp scanning
```

```nasm
;字符串不相同，打印
not_equal2:
      mov rdi, string22
      mov rsi, rax
      xor  rax,rax
      call printf
; 扫描字符串中的一个字符--------------------------------
; 首先打印字符串
      mov rdi,string33
      mov rsi,string3
      xor  rax,rax
      call printf
; 然后打印搜索参数，只能是 1 个字符
      mov rdi,string46
      mov rsi,string4
      xor  rax,rax
      call printf
scanning:
      lea  rdi,[string3]     ; 字符串
      lea  rsi,[string4]     ; 搜索参数
      mov rdx, strlen3
      call cscan
      cmp rax,0
      jz  char_not_found
;找到字符，打印
      mov rdi,string44
      mov rsi,string4
      mov rdx,rax
      xor rax,rax
      call printf
      jmp exit
;找不到字符，打印
char_not_found:
      mov rdi,string45
      mov rsi,string4
      xor rax,rax
      call printf
exit:
leave
```

```
            ret
; 函数================================================

; 用于比较两个字符串的函数----------------------------------
  compare1:   mov   rcx, rdx
              cld
  cmpr:       cmpsb
              jne   notequal
              loop  cmpr
              xor   rax,rax
              ret
  notequal:   mov   rax, strlen2
              dec   rcx          ;计算位置
              sub   rax,rcx      ;计算位置
              ret
              xor   rax,rax
              ret
;-----------------------------------------------------------------
; 用于比较两个字符串的函数----------------------------------
  compare2:   mov   rcx, rdx
              cld
              repe cmpsb
              je    equal2
              mov   rax, strlen2
              sub   rax,rcx      ;计算位置
              ret
  equal2:     xor   rax,rax
              ret
;-----------------------------------------------------------------
;函数用于从一个字符串中查找一个字符
  cscan:      mov   rcx, rdx
              lodsb
              cld
              repne scasb
              jne   char_notfound
              mov   rax, strlen3
              sub   rax,rcx      ;计算位置
              ret
  char_notfound:    xor rax,rax
```

```
    ret
```

为了进行比较，我们将讨论两个版本。和以前一样，我们将第一个(源)字符串的地址放入 rsi，将第二个字符串(目标)的地址放入 rdi，并将字符串长度放入 rcx。为了确认，我们用 cld 清除了方向标志 DF。因此，我们向前迈进了一步。

指令 cmpsb 比较两个字节，如果两个字节相等，则将状态标志 ZF 设置为 1；如果两个字节不相等，则将状态标志 ZF 设置为 0。

使用 ZF 标志可能会造成困惑。如果 ZF=1，则意味着刚执行的指令的结果为 0(字节相等)。如果 ZF=0，则意味着刚执行的指令的结果不为 0(字节不相等)。因此，我们必须找出 ZF 是否为 0 以及何时变为 0。为测试 ZF 并根据测试结果继续执行，我们有许多跳转指令，如下所示。

- jz: Jump if zero (ZF=1)
  - 与 je 等效：相等时跳转(ZF=1)(字节相等)
- jnz: Jump if not zero (ZF=0)
  - 与 jne 等效：如果不相等则跳转(ZF=0)(字节不相等)

当未设置 DF 时，寄存器 rsi 和 rdi 由 cmpsb 增加，在设置 DF 时由 cmpsb 减少。我们创建一个执行 cmpsb 的循环，直到 ZF 变为 0。当 ZF 变为 0 时，执行跳出循环并开始根据 rcx 中的值计算不同字符的位置。但是，rcx 仅在循环结束时才进行调整，而循环从未完成，因此我们必须调整 rcx(将其减小 1)。结果位置返回到 rax 中的 main。

在用于比较的第二个版本中，我们将使用 repe，它是 rep 的一个版本，意思是"相等时重复"。与以前一样，cmpsb 根据比较设置 ZF，ZF=1 表示字节相等。一旦 cmpsb 将 ZF 设置为 0，repe 循环就结束了，rcx 可以用来计算不同字符出现的位置。如果字符串完全相同，则 rcx 将为 0，ZF 将为 1。重复执行后，指令 je 会测试 ZF 是否等于 1。如果 ZF 为 1，则字符串相等；否则，字符串等于 0。如果为 0，则字符串不相等。我们使用 rcx 来计算不同的位置，因此不必调整 rcx，因为 repe 在每个循环中都会先减少 rcx。

扫描的工作原理类似，使用 repne，表示"重复但不相等"。我们还使用 lodsb 并将地址 rsi 处的字节加载到 rax 中。指令 scasb 将 al 中的字节(rax 中的低字节)与 rdi 指向的字节进行比较，并相应地设置(1 表示相等)或复位(0 表示不相等)ZF 标志。指令 repne 查看状态标志，如果 ZF=0，则继续；也就是说，两个字节不相等。如果两个字节相等，scasb 将 ZF 设置为 1，repne 循环停止，rcx 可用于计算字节在字符串中的位置。

扫描只能使用一个字符作为搜索参数；如果你想知道如何使用字符串作为搜索参数，则必须逐个字符进行扫描。

输出如图 24-2 所示。

```
jo@UbuntuDesktop:~/Desktop/linux64/gcc/31 strings 2$ make
nasm -f elf64 -g -F dwarf strings.asm -l strings.lst
gcc -o strings strings.o
jo@UbuntuDesktop:~/Desktop/linux64/gcc/31 strings 2$ ./strings
This is the 1st string.
This is the 2nd string.
Comparing strings: The strings differ, starting at position: 13.
Now look at this string: The quick brown fox jumps over the lazy dog.
Scanning for the character 'z'.
The character 'z' was found at position: 38.
jo@UbuntuDesktop:~/Desktop/linux64/gcc/31 strings 2$
```

图 24-2 strings.asm 的输出

24.3 小结

本章内容：
- 以极其高效的方式移动和复制内存块
- 使用 movsb 和 rep
- 比较和扫描内存块
- 使用 cmpsb、scasb、repe 和 repne

第 25 章

cpuid

有时找出处理器的可用功能也很必要。例如,在你的程序中,可以查找是否存在特定版本的 SSE。在下一章中,我们将使用带有 SSE 指令的程序,因此我们需要首先找出处理器支持的 SSE 版本。有一条检查 CPU 特性的指令:cpuid。

25.1 使用 cpuid

首先将一个特定参数放在 eax 中,然后执行指令 cpuid,最后在 ecx 和 edx 中检查返回的值。确实,cpuid 使用 32 位寄存器。

使用 cpuid 可以发现的信息量是惊人的。参考英特尔手册(https://software.intel.com/sites/default/files/managed/39/c5/32542-sdm-vol-1-2abcd-3abcd.pdf)并在第 2A 卷中查找 cpuid 指令。当你在 eax 中用某个值启动 cpuid 时,你会发现有几个表显示了在 ecx 中返回的内容。这只是可以检索的信息的一部分;另一个表显示了在 edx 中返回的信息。浏览英特尔手册,了解各种可能的情况。

下面看一个在下一章中需要使用的查找 SSE 功能的示例。在英特尔手册中,你可以使用 ecx 的第 0、19 和 20 位以及 edx 的第 25 和 26 位来找出在处理器中实现的 SSE 版本。

代码清单 25-1 显示了示例程序。

代码清单 25-1:cpu.asm

```
; cpu.asm
extern printf
section .data
        fmt_no_sse db "This cpu does not support SSE",10,0
        fmt_sse42 db "This cpu supports SSE 4.2",10,0
```

```asm
            fmt_sse41 db "This cpu supports SSE 4.1",10,0
            fmt_ssse3 db "This cpu supports SSSE 3",10,0
            fmt_sse3 db "This cpu supports SSE 3",10,0
            fmt_sse2 db "This cpu supports SSE 2",10,0
            fmt_sse db "This cpu supports SSE",10,0
section .bss
section .text
        global main
main:
push rbp
mov   rbp,rsp
    call cpu_sse        ;如果支持 sse,则在 rax 中返回 1,否则返回 0
leave
ret

cpu_sse:
        push rbp
        mov rbp,rsp
        xor r12,r12         ;SSE 标志可用
        mov eax,1           ;请求 CPU 功能标志
        cpuid

    ;SSE 测试
        test edx,2000000h   ;测试 bit 25 (SSE)
        jz sse2             ;SSE 可用
        mov r12,1
        xor rax,rax
        mov rdi,fmt_sse
        push rcx            ;由 printf 修改
        push rdx            ;cpuid 的保留结果
        call printf
        pop rdx
        pop rcx
    sse2:
        test edx,4000000h   ;测试 bit 26 (SSE 2)
        jz sse3             ;SSE 2可用
```

第 25 章 cpuid

```
        mov r12,1
        xor rax,rax
        mov rdi,fmt_sse2
        push rcx              ;由 printf 修改
        push rdx              ;cpuid 的保留结果
        call printf
        pop rdx
        pop rcx
sse3:
        test ecx,1            ;测试 bit 0 (SSE 3)
        jz ssse3              ;SSE 3 可用
        mov r12,1
        xor rax,rax
        mov rdi,fmt_sse3
        push rcx              ;由 printf 修改
        call printf
        pop rcx
ssse3:
        test ecx,9h           ;测试 bit 0 (SSE 3)
        jz sse41              ;SSE 3 可用
        mov r12,1
        xor rax,rax
        mov rdi,fmt_ssse3
        push rcx              ;由 printf 修改
        call printf
        pop rcx
sse41:
        test ecx,80000h       ;测试 bit 19 (SSE 4.1)
        jz sse42              ;SSE 4.1 可用
        mov r12,1
        xor rax,rax
        mov rdi,fmt_sse41
        push rcx              ;由 printf 修改
        call printf
        pop rcx
sse42:
```

```
    test ecx,100000h       ;测试 bit 20 (SSE 4.2)
        jz wrapup          ;SSE 4.2 可用
        mov r12,1
        xor rax,rax
        mov rdi,fmt_sse42
        push rcx           ;由 printf 修改
        call printf
        pop rcx
    wrapup:
        cmp r12,1
        je sse_ok
        mov rdi,fmt_no_sse
        xor rax,rax
        call printf        ;如果 SSE 不可用,则显示消息
        jmp the_exit
    sse_ok:
        mov rax,r12        ;返回 1,支持 sse
    the_exit:
    leave
    ret
```

主程序仅调用一个函数 cpu_sse,如果返回值为 1,则处理器支持某些版本的 SSE。如果返回值为 0,则该计算机不支持 SSE。在函数 cpu_sse 中,我们找出支持哪些 SSE 版本。将 1 放入 eax 并执行指令 cupid;如前所述,结果将在 ecx 和 edx 中返回。

25.2 使用 test 指令

ecx 和 edx 寄存器将使用一条测试指令进行求值,该指令是两个操作数的按位逻辑"与"。我们本可使用 cmp 指令,但 test 具有性能优势。当然,也可以使用 bt 指令(请参阅第 17 章)。

test 指令根据测试结果设置标志 SF、ZF 和 PF。在英特尔手册中,你将看到 test 指令的操作,如下所示:

```
TEMP ← SRC1 AND SRC2;
SF ← MSB(TEMP);
IF TEMP = 0
```

```
            THEN ZF ← 1;
            ELSE ZF ← 0;
    FI:
    PF ← BitwiseXNOR(TEMP[0:7]);
    CF ← 0;
    OF ← 0;
    (* AF is undefined *)
```

在我们的例子中，重要标志是 ZF。如果 ZF=0，则结果为 nonzero(非零)；SSE 位为 1，CPU 支持该版本的 SSE。指令 jz 评估是否 ZF=1，如果是，则不支持 SSE 版本，执行跳转到下一节。否则，程序将打印一条确认消息。

在我们的示例中，执行 cpuid 之后测试 edx。寄存器 edx 具有 32 个位，我们想知道第 25 位(bit 25)是否已设置，这意味着 CPU 支持 SSE(版本 1)。因此，我们需要 test 指令中的第二个操作数的第 25 位是 1，其他位均为 0。请记住，最低位的索引为 0，最高位的索引为 31。它的二进制形式看起来像这样：

```
0000 0010 0000 0000 0000 0000 0000 0000
```

它的十六进制形式如下所示：

```
2000000
```

请记住，你可以在互联网上找到许多从二进制到十六进制的转换工具。

通过程序执行"级联"，如果不支持 SSE，则 r12 将保持为 0。我们没有使用返回值，但是你可以检查返回值 rax 以得出是否支持任何 SSE 的结论。或者，可以修改程序以返回最高版本的 SSE。

图 25-1 显示了输出。

```
jo@UbuntuDesktop:~/Desktop/linux64/gcc/32 cpu_sse$ make
nasm -f elf64 -g -F dwarf cpu_sse.asm -l cpu_sse.lst
gcc -o cpu_sse cpu_sse.o
jo@UbuntuDesktop:~/Desktop/linux64/gcc/32 cpu_sse$ ./cpu_sse
This cpu supports SSE.
This cpu supports SSE 2.
This cpu supports SSE 3.
This cpu supports SSSE 3.
This cpu supports SSE 4.1.
This cpu supports SSE 4.2.
jo@UbuntuDesktop:~/Desktop/linux64/gcc/32 cpu_sse$
```

图 25-1 cpu_sse.asm 的输出

可以构建类似的功能来查找其他 CPU 信息，并根据返回的结果选择使用此 CPU 上的某些功能还是使用另一个 CPU 上的其他功能。

在下一章讨论 AVX 时，我们将不得不再次确定 CPU 是否支持 AVX。

25.3　小结

本章内容：
- 如何用 cpuid 找出 CPU 支持哪些功能
- 如何在 test 指令中使用位(bit)

第 26 章

SIMD

SIMD 是单指令流、多数据流的缩写。SIMD 是 Michael J. Flynn 提出的一个术语，指的是允许你在多个数据"流"上执行一条指令的功能。SIMD 可以潜在地提高程序的性能，是并行计算的一种形式。但某些情况下，取决于硬件功能和要执行的指令，不同数据流上的执行可能会顺序发生。你可以在此处找到有关 Flynn 分类的更多信息：

https://ieeexplore.ieee.org/document/5009071/

以及这里：

https://en.wikipedia.org/wiki/Flynn%27s_taxonomy

SIMD 的第一个实现是 MMX，似乎没有人知道 MMX 的确切含义。它可能是指多媒体扩展、多重数学扩展或矩阵数学扩展。无论如何，MMX 被数据流 SIMD 扩展(Streaming SIMD Extension，SSE)指令集取代。后来的 SSE 由 AVX(Advanced Vector Extension，高级向量扩展)进行了扩展。在这里，我们将以 SSE 作为基础进行介绍，我们将在下一章介绍 AVX。

26.1 标量数据和打包数据

支持 SSE 功能的处理器有 16 个额外的 128 位寄存器(xmm0 到 xmm15)和一个控制寄存器 mxcsr。我们已经使用 xmm 寄存器来执行浮点计算，但是我们可以用这些高级寄存器做更多事情。xmm 寄存器可以包含标量数据或打包数据。

对于标量数据，指的是一个值。当我们将 3.141592654 放在 xmm0 中时，则 xmm0 包含一个标量值。也可以在 xmm0 中存储多个值。这些值称为打包数据。以下是可能在 xmm 寄存器中存储的值：
- 两个 64 位双精度浮点数
- 四个 32 位单精度浮点数
- 两个 64 位整数(quadwords)
- 四个 32 位整数(双字)
- 八个 16 位短整数(字)
- 16 个 8 位字节或字符

示意图如图 26-1 所示。

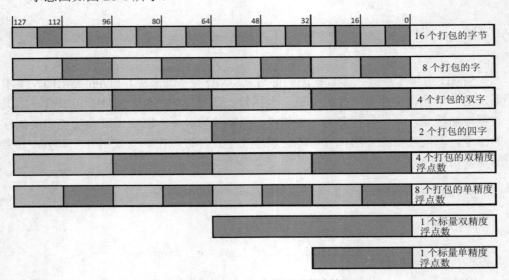

图 16-1　一个 xmm 寄存器的内容

标量数字和打包数字有不同的汇编指令。在英特尔手册中，你可以看到有大量的 SSE 指令可用。在本章和后续各章中，我们将介绍几个示例作为入门，以助你一臂之力。

我们将在后续章节中使用 AVX 功能。AVX 寄存器的大小是 xmm 的两倍。AVX 寄存器称为 ymm 寄存器，有 256 个位。还有 AVX-512，它提供具有 512 个位的 AVX-512 寄存器，称为 zmm 寄存器。

由于具有并行计算的潜力，因此 SIMD 可以在图像处理、音频处理、信号处理、矢量和矩阵运算等广泛的应用中提高计算速度。在后续章节中，我们将使用 SIMD 进行矩阵操作。请放心，我们将把对数学知识的需要限制在基本矩阵运算上。我们的目的是学习 SIMD，而不是线性代数。

26.2 数据对齐与不对齐

内存中的数据可以不对齐，也可以在 16、32 等倍数的某些地址上对齐。在内存中对齐数据可以大大提高程序的性能。原因是对齐的压缩 SSE 指令希望同时获取 16 个字节的内存块。参见图 26-2 的左侧。当内存中的数据没有对齐时，CPU 必须执行多个获取操作来得到所需的 16 字节数据，这会降低执行速度。我们有两种类型的 SSE 指令：对齐的指令和未对齐的打包指令。未对齐的打包指令可以处理未对齐的内存，但是通常存在性能劣势。

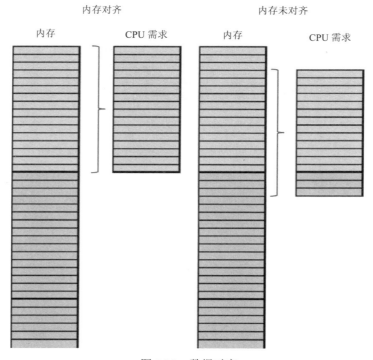

图 26-2　数据对齐

当使用 SSE 时，对齐意味着 section.data 和 section.bss 中的数据应该在 16 字节的边界上对齐。在 NASM 中，可以在要对齐的数据前使用汇编指令 align 16 和 alignb 16。你将在接下来的章节中看到这样的示例。对于 AVX，数据应在 32 字节的边界上对齐，对于 AVX-512，数据应在 64 位的边界上对齐。

26.3 小结

本章内容：
- SSE 提供 16 个额外的 128 位寄存器
- 标量数据和打包数据之间的区别
- 数据对齐的重要性

第 27 章

小心 mxcsr

在深入研究 SSE 编程之前，你需要了解用于浮点操作的 SSE 控制和状态寄存器——mxcsr。它是一个 32 位寄存器，只使用较低的 16 位。布局如表 27-1 所示。

表 27-1 mxcsr 的布局

位(Bit)	助记符	意义
0	IE	无效的操作错误(Invalid operation error)
1	DE	非规格化错误(Denormal error)
2	ZE	被零除错误(Divide-by-zero)
3	OE	溢出错误(Overflow error)
4	UE	下溢错误(Underflow error)
5	PE	精度错误(Precision error)
6	DAZ	非规格化浮点数置零(Denormals are zeros)
7	IM	无效的操作掩码(Invalid operation mask)
8	DM	非规格化操作掩码(Denormal operation mask)
9	ZM	被零除掩码(Divide-by-zero mask)
10	OM	溢出掩码(Overflow mask)
11	UM	下溢掩码(Underflow mask)
12	PM	精度掩码(Precision mask)
13	RC	舍入控制(Rounding control)
14	RC	舍入控制(Rounding control)
15	FZ	下溢则零(Flush to zero)

位 0 到 5 表示何时检测到浮点异常，例如被零除，或者由于浮点运算而导致值失去一些精度。位 7 到 12 是掩码，控制浮点操作在位 0 到 5 中设置标志时的行为。

例如，如果发生被零除的情况，程序通常会抛出错误并可能崩溃。当你将被零除掩码标记设置为 1 时，程序将不会崩溃，并且你可以执行某些指令来减缓崩溃。默认情况下掩码设置为 1，因此不会引发 SIMD 浮点异常。使用两个位(位 13 和 14)来控制舍入，如表 27-2 所示。

表 27-2 控制舍入

位(Bits)	意义
00	四舍五入到最接近(Round to nearest)
01	向下取整(Round down)
10	向上取整(Round up)
11	截断(Truncate)

我们不会讨论 mxcsr 寄存器的所有状态和掩码详细信息。要了解更多信息，请参阅英特尔手册。

27.1 操作 mxcsr 的位

可以使用 ldmxcsr 和 stmxcsr 指令来操作 mxcsr 寄存器中的位。默认的 mxcsr 状态为 00001F80 或 0001 1111 1000 0000。设置所有掩码位，舍入设置为最接近。

代码清单 27-1 到代码清单 27-4 展示了使用 mxcsr 可以做什么的示例。

代码清单 27-1：mxcsr.asm

```
; mxcsr.asm
extern printf
extern print_mxcsr
extern print_hex
section .data
        eleven    dq 11.0
        two       dq 2.0
        three     dq 3.0
        ten       dq 10.0
        zero      dq 0.0
        hex       db "0x",0
        fmt1      db 10,"Divide, default mxcsr:",10,0
        fmt2      db 10,"Divide by zero, default mxcsr:",10,0
        fmt4      db 10,"Divide, round up:",10,0
```

```
        fmt5        db 10,"Divide, round down:",10,0
        fmt6        db 10,"Divide, truncate:",10,0
        f_div       db "%.1f divided by %.1f is %.16f, in hex: ",0
        f_before    db 10,"mxcsr before:",9,0
        f_after     db "mxcsr after:",9,0
;mxcsr 值
        default_mxcsr   dd 0001111110000000b
        round_nearest   dd 0001111110000000b
        round_down      dd 0011111110000000b
        round_up        dd 0101111110000000b
        truncate        dd 0111111110000000b
section .bss
        mxcsr_before    resd 1
        mxcsr_after     resd 1
        xmm     resq 1
section .text
        global main
main:
push rbp
mov rbp,rsp

;除法
;默认 mxcsr
        mov     rdi,fmt1
        mov     rsi,ten
        mov     rdx,two
        mov     ecx,[default_mxcsr]
        call    apply_mxcsr
;-----------------------------------------
;除法(精度误差)
;默认 mxcsr
        mov     rdi,fmt1
        mov     rsi,ten
        mov     rdx,three
        mov     ecx,[default_mxcsr]
        call    apply_mxcsr
```

```
;除以零
;默认mxcsr
        mov   rdi,fmt2
        mov   rsi,ten
        mov   rdx,zero
        mov   ecx, [default_mxcsr]
        call  apply_mxcsr
;除法(精度误差)
;向上取整
        mov   rdi,fmt4
        mov   rsi,ten
        mov   rdx,three
        mov   ecx, [round_up]
        call  apply_mxcsr
;除法(精度误差)
;向上取整
        mov   rdi,fmt5
        mov   rsi,ten
        mov   rdx,three
        mov   ecx, [round_down]
        call  apply_mxcsr
;除法(精度误差)
;截断
        mov   rdi,fmt6
        mov   rsi,ten
        mov   rdx,three
        mov   ecx, [truncate]
        call  apply_mxcsr
;----------------------------------------------
;除法(精度误差)
;默认mxcsr
        mov   rdi,fmt1
        mov   rsi,eleven
        mov   rdx,three
        mov   ecx, [default_mxcsr]
        call  apply_mxcsr       ;除法(精度误差)
```

```asm
;向上取整
        mov rdi,fmt4
        mov rsi,eleven
        mov rdx,three
        mov ecx, [round_up]
        call apply_mxcsr
;除法(精度误差)
;向上取整
        mov rdi,fmt5
        mov rsi,eleven
        mov rdx,three
        mov ecx, [round_down]
        call apply_mxcsr
;除法(精度误差)
;截断
        mov rdi,fmt6
        mov rsi,eleven
        mov rdx,three
        mov ecx, [truncate]
        call apply_mxcsr
leave
ret

;函数 ------------------------------------------
apply_mxcsr:
push    rbp
mov     rbp,rsp
        push    rsi
        push    rdx
        push    rcx
        push    rbp         ; 另一个用于堆栈对齐
        call    printf
        pop     rbp
        pop     rcx
        pop     rdx
        pop     rsi

        mov     [mxcsr_before],ecx
```

```
        ldmxcsr   [mxcsr_before]
        movsd     xmm2, [rsi]        ; 双精度浮点数进入 xmm2
        divsd     xmm2, [rdx]        ; 除以 xmm2
        stmxcsr   [mxcsr_after]      ; 将 mxcsr 保存到内存
        movsd     [xmm],xmm2         ; 供 print_xmm 使用
        mov       rdi,f_div
        movsd     xmm0, [rsi]
        movsd     xmm1, [rdx]
        call      printf
        call      print_xmm
;打印 mxcsr
        mov       rdi,f_before
        call      printf
        mov       rdi, [mxcsr_before]
        call      print_mxcsr
        mov       rdi,f_after
        call      printf
        mov       rdi, [mxcsr_after]
        call      print_mxcsr
leave
ret
;函数 -------------------------------------------
print_xmm:
push  rbp
mov   rbp,rsp
        mov  rdi, hex        ;打印 0x
        call printf
        mov  rcx,8
.loop:
        xor  rdi,rdi
        mov  dil,[xmm+rcx-1]
        push rcx
        call print_hex
        pop  rcx
        loop .loop
leave
```

```
        ret
```

代码清单 27-2：print_hex.c

```c
// print_hex.c

#include <stdio.h>

void print_hex(unsigned char n){
        if (n < 16) printf("0");
        printf("%x",n);
}
```

代码清单 27-3：print_mxcsr.c

```c
// print_mxcsr.c

#include <stdio.h>

void print_mxcsr(long long n){
    long long s,c;
    for (c = 15; c >= 0; c--)
    {
        s = n >> c;
        // 每 8 位后的空间
        if ((c+1) % 4 == 0) printf(" ");
        if (s & 1)
                printf("1");
        else
                printf("0");
    }
    printf("\n");
}
```

代码清单 27-4：makefile

```
mxcsr: mxcsr.o print_mxcsr.o print_hex.o
    gcc -o mxcsr mxcsr.o print_mxcsr.o print_hex.o -no-pie
mxcsr.o: mxcsr.asm
    nasm -f elf64 -g -F dwarf mxcsr.asm -l mxcsr.lst
print_mxcsr: print_mxcsr.c
```

```
    gcc -c print_mxcsr.c
print_hex: print_hex.c
    gcc -c print_hex.c
```

在这个程序中，我们展示了不同的舍入模式和一个掩码零除法。默认为舍入到最近值。例如，在十进制中，计算以.5 或更大结尾的正数将舍入到更高的数字，而以.5 或更大结尾的负数将舍入到更低(更负)的数字。但是，这里是十六进制(而不是十进制)的四舍五入，并且得出的结果未必与十进制的四舍五入相同！

输出如图 27-1 所示。

```
jo@UbuntuDesktop:~/Desktop/linux64/gcc/34 mxcsr$ ./mxcsr
Divide, default mxcsr:
10.0 divided by 2.0 is 5.0000000000000000, in hex: 0x4014000000000000
mxcsr before:    0001 1111 1000 0000
mxcsr after:     0001 1111 1000 0000

Divide, default mxcsr:
10.0 divided by 3.0 is 3.3333333333333335, in hex: 0x400aaaaaaaaaaaab
mxcsr before:    0001 1111 1000 0000
mxcsr after:     0001 1111 1010 0000

Divide by zero, default mxcsr:
10.0 divided by 0.0 is inf, in hex: 0x7ff0000000000000
mxcsr before:    0001 1111 1000 0000
mxcsr after:     0001 1111 1000 0100

Divide, round up:
10.0 divided by 3.0 is 3.3333333333333335, in hex: 0x400aaaaaaaaaaaab
mxcsr before:    0101 1111 1000 0000
mxcsr after:     0101 1111 1010 0000

Divide, round down:
10.0 divided by 3.0 is 3.3333333333333330, in hex: 0x400aaaaaaaaaaaaa
mxcsr before:    0011 1111 1000 0000
mxcsr after:     0011 1111 1010 0000

Divide, truncate:
10.0 divided by 3.0 is 3.3333333333333330, in hex: 0x400aaaaaaaaaaaaa
mxcsr before:    0111 1111 1000 0000
mxcsr after:     0111 1111 1010 0000

Divide, default mxcsr:
11.0 divided by 3.0 is 3.6666666666666665, in hex: 0x400d555555555555
mxcsr before:    0001 1111 1000 0000
mxcsr after:     0001 1111 1010 0000

Divide, round up:
11.0 divided by 3.0 is 3.6666666666666670, in hex: 0x400d555555555556
mxcsr before:    0101 1111 1000 0000
mxcsr after:     0101 1111 1010 0000

Divide, round down:
11.0 divided by 3.0 is 3.6666666666666665, in hex: 0x400d555555555555
mxcsr before:    0011 1111 1000 0000
mxcsr after:     0011 1111 1010 0000

Divide, truncate:
11.0 divided by 3.0 is 3.6666666666666665, in hex: 0x400d555555555555
mxcsr before:    0111 1111 1000 0000
mxcsr after:     0111 1111 1010 0000
jo@UbuntuDesktop:~/Desktop/linux64/gcc/34 mxcsr$
```

图 27-1　mxcsr.asm 的输出

27.2 分析程序

下面分析一下程序。我们对多个除法运算的结果进行四舍五入。除法运算是在 apply_mxcsr 函数中完成的。在调用此函数前，我们将打印标题的地址放入 rdi，将被除数放入 rdi，将除数放入 rdx。然后将所需的 mxcsr 值从内存复制到 ecx；对于第一个调用，它是默认的 mxcsr 值。然后我们调用 apply_mxcsr。在这个函数中，我们打印标题，不要忘记首先保留必要的寄存器并对齐堆栈。然后，我们将 ecx 中的值存储到 mxcsr_before，并使用 ldmxcsr 指令将 mxcsr_before 中存储的值加载到 mxcsr 中。指令 ldmxcsr 采用 32 位内存变量(双字)作为操作数。指令 divsd 将 xmm 寄存器作为第一个参数，并将 xmm 寄存器或 64 位变量作为第二个操作数。完成除法后，使用 stmxcsr 指令将 mxcsr 寄存器的内容存储到变量 mxcsr_after 的内存中。我们将 xmm2 中的商复制到变量 xmm 的内存中以进行打印。

我们先以十进制形式打印商，然后在同一行上以十六进制形式打印商。我们不能在汇编中使用 printf 打印十六进制值(至少不能在此处使用的版本进行打印)；我们必须为此创建一个函数。因此，我们创建了函数 print_xmm。此函数接收内存变量 xmm 并在循环中将字节逐个加载到 dil 中。在同一循环中，将为每个字节调用定制的 C 函数 print_hex。通过在地址中使用递减循环计数器 rcx，我们还可以处理低字节序：浮点值以低字节序格式存储在内存中！

最后，显示 mxcsr_before 和 mxcsr_after 以便我们可以进行比较。函数 print_mxcsr 用于打印 mxcsr 中的位，并且与我们在前几章中使用的位打印函数相似。

有些读者可能会觉得这很复杂，使用调试器逐步执行程序，并观察内存和寄存器。

分析一下输出可以发现，将 10 除以 2 时，mxcsr 不会改变。将 10 除以 3 时，可得到 3.333。这里 mxcsr 在位 5 中发出精度错误信号。默认舍入(舍入到最接近的值)将最后一个十六进制从 a 增加到 b。在十进制中，是向下舍入；但是在十六进制中，大于 8 的 a 将向上舍入到 b。

我们继续进行零除法：mxcsr 在位 2 发出零除法信号，但是由于设置了零除法掩码 ZE，所以程序不会崩溃。结果为 inf 或 0x7ff0000000000000。

下一个除法和向上舍入的结果与舍入到最接近值的结果相同。接下来的两个除法是舍入和截断，结果是一个最后数位是 a 的十六进制数字。

为显示舍入的差异，我们用 11 除以 3 进行相同的练习。这个除法会得到一个最后数位(十六进制)比较低的商。你可以比较舍入行为。

作为练习，将除以零掩码位清除并重新运行程序。你将看到程序会崩溃。除以零掩码和其他掩码允许你捕获错误并跳转到某个错误过程。

27.3　小结

本章内容：
- mxcsr 寄存器的设计和用途
- 如何操作 mxcsr 寄存器
- 如何处理舍入

第 28 章

SSE 对齐

是时候开始真正的 SSE 工作了！尽管有多个章节介绍 SSE，但我们只是触及了这个主题的表面。有数百条 SIMD 指令(MMX、SSE、AVX)，深入研究它们需要另一本书甚至一系列书。本章将给出一些示例，以便你知道从哪里开始。这些示例的目的是让你能够在英特尔手册里的大量 SIMD 指令中找到自己想要的。本章将讨论对齐，第 26 章曾简要介绍过对齐。

28.1 未对齐示例

代码清单 28-1 显示了如何使用内存中未对齐的数据添加向量。

代码清单 28-1：sse_unaligned.asm

```
; sse_unaligned.asm
extern printf
section .data
;单精度
        spvector1   dd  1.1
                    dd  2.2
                    dd  3.3
                    dd  4.4
        spvector2   dd  1.1
                    dd  2.2
                    dd  3.3
                    dd  4.4
;双精度
        dpvector1   dq  1.1
```

```nasm
                dq 2.2
        dpvector2   dq 3.3
                dq 4.4

        fmt1 db "Single Precision Vector 1: %f, %f, %f, %f",10,0
        fmt2 db "Single Precision Vector 2: %f, %f, %f, %f",10,0
        fmt3 db "Sum of Single Precision Vector 1 and Vector 2:"
             db " %f, %f, %f, %f",10,0
        fmt4 db "Double Precision Vector 1: %f, %f",10,0
        fmt5 db "Double Precision Vector 2: %f, %f",10,0
        fmt6 db "Sum of Double Precision Vector 1 and Vector 2:"
             db " %f, %f",10,0

section .bss
        spvector_res    resd 4
        dpvector_res    resq 4
section .text
        global main
main:
push rbp
mov rbp,rsp

; 添加两个单精度浮点向量
        mov rsi,spvector1
        mov rdi,fmt1
        call printspfp

        mov rsi,spvector2
        mov rdi,fmt2
        call printspfp

        movups  xmm0, [spvector1]
        movups  xmm1, [spvector2]
        addps   xmm0,xmm1
        movups  [spvector_res], xmm0
        mov rsi,spvector_res
        mov rdi,fmt3
        call printspfp

; 添加两个双精度浮点向量
        mov     rsi,dpvector1
```

```
            mov     rdi,fmt4
            call    printdpfp

            mov     rsi,dpvector2
            mov     rdi,fmt5
            call    printdpfp

            movupd  xmm0, [dpvector1]
            movupd  xmm1, [dpvector2]
            addpd   xmm0,xmm1
            movupd  [dpvector_res], xmm0
            mov     rsi,dpvector_res
            mov     rdi,fmt6
            call    printdpfp
leave
ret

printspfp:
push    rbp
mov     rbp,rsp
            movss     xmm0, [rsi]
            cvtss2sd  xmm0,xmm0
            movss     xmm1, [rsi+4]
            cvtss2sd  xmm1,xmm1
            movss     xmm2, [rsi+8]
            cvtss2sd  xmm2,xmm2
            movss     xmm3, [rsi+12]
            cvtss2sd  xmm3,xmm3
            mov       rax,4       ; 四个浮点数
            call      printf
leave
ret
printdpfp:
push    rbp
mov     rbp,rsp
            movsd   xmm0, [rsi]
            movsd   xmm1, [rsi+8]
            mov rax,2       ; 四个浮点数
            call printf
leave
```

```
ret
```

第一个 SSE 指令是 movups(意思是"移动未对齐的打包单精度"），它将数据从内存复制到 xmm0 和 xmm1 中。结果，xmm0 和 xmm1 各包含一个具有四个单精度值的向量。然后，我们使用 addps(意思是"打包单精度数据相加")将两个向量相加；所得向量放入 xmm0，然后传输到内存。然后，我们使用函数 printspfp 打印结果。在 printspfp 函数中，我们使用 movss(意思是"移动标量单精度")将内存中的每个值复制到 xmm 寄存器中。因为 printf 需要双精度浮点参数，所以我们使用指令 cvtss2sd(意思是"将标量单精度转换为标量双精度")将单精度浮点数转换为双精度。

接下来，我们添加两个双精度值。这个过程类似于添加单精度数字，但是我们使用 movupd 和 addpd 来实现双精度。用于打印双精度的 printdpfp 函数要简单一些。我们只有一个二元向量，并且因为已经在使用双精度，所以不必转换向量。

输出如图 28-1 所示。

```
jo@UbuntuDesktop:~/Desktop/linux64/gcc/33 sse_unaligned$ make
nasm -f elf64 -g -F dwarf sse_unaligned.asm -l sse_unaligned.lst
gcc -o sse_unaligned sse_unaligned.o
jo@UbuntuDesktop:~/Desktop/linux64/gcc/33 sse_unaligned$ ./sse_unaligned
Single Precision Vector 1: 1.100000, 2.200000, 3.300000, 4.400000
Single Precision Vector 2: 1.100000, 2.200000, 3.300000, 4.400000
Sum of Single Precision Vector 1 and Vector 2: 2.200000, 4.400000, 6.600000, 8.800000
Double Precision Vector 1: 1.100000, 2.200000
Double Precision Vector 2: 3.300000, 4.400000
Sum of Double Precision Vector 1 and Vector 2: 4.400000, 6.600000
jo@UbuntuDesktop:~/Desktop/linux64/gcc/33 sse_unaligned$
```

图 28-1　sse_unaligned.asm 的输出

28.2　对齐示例

代码清单 28-2 展示了如何将两个向量相加。

代码清单 28-2：sse_aligned.asm

```
; sse_aligned.asm
extern printf
section .data
        dummy   db 13
align 16
        spvector1   dd 1.1
                    dd 2.2
                    dd 3.3
                    dd 4.4
        spvector2   dd 1.1
```

```nasm
                dd 2.2
                dd 3.3
                dd 4.4
        dpvector1   dq 1.1
                    dq 2.2
        dpvector2   dq 3.3
                    dq 4.4

        fmt1 db "Single Precision Vector 1: %f, %f, %f, %f",10,0
        fmt2 db "Single Precision Vector 2: %f, %f, %f, %f",10,0
        fmt3 db "Sum of Single Precision Vector 1 and Vector 2:"
             db " %f, %f, %f, %f",10,0
        fmt4 db "Double Precision Vector 1: %f, %f",10,0
        fmt5 db "Double Precision Vector 2: %f, %f",10,0
        fmt6 db "Sum of Double Precision Vector 1 and Vector 2:"
             db " %f, %f",10,0
section .bss
alignb 16
        spvector_res    resd 4
        dpvector_res    resq 4
section .text
        global main
main:
push rbp
mov rbp,rsp

; 添加两个单精度浮点向量
        mov     rsi,spvector1
        mov     rdi,fmt1
        call    printspfp

        mov     rsi,spvector2
        mov     rdi,fmt2
        call    printspfp

        movaps  xmm0, [spvector1]
        addps   xmm0, [spvector2]

        movaps  [spvector_res], xmm0
        mov     rsi,spvector_res
```

```
        mov     rdi,fmt3
        call    printspfp

; 添加两个双精度浮点向量
        mov     rsi,dpvector1
        mov     rdi,fmt4
        call    printdpfp

        mov     rsi,dpvector2
        mov     rdi,fmt5
        call    printdpfp

        movapd  xmm0, [dpvector1]
        addpd   xmm0, [dpvector2]

        movapd  [dpvector_res], xmm0
        mov     rsi,dpvector_res
        mov     rdi,fmt6
        call    printdpfp
; exit
mov rsp,rbp
pop rbp         ; 撤消开始时的压入
ret

printspfp:
push   rbp
mov    rbp,rsp
        movss     xmm0, [rsi]
        cvtss2sd  xmm0,xmm0    ;printf 需要双精度参数
        movss     xmm1, [rsi+4]
        cvtss2sd  xmm1,xmm1
        movss     xmm2, [rsi+8]
        cvtss2sd  xmm2,xmm2
        movss     xmm3, [rsi+12]
        cvtss2sd  xmm3,xmm3
        mov       rax,4      ; 四个浮点数
        call printf
leave
ret

printdpfp:
```

```
        push    rbp
        mov     rbp,rsp
                movsd   xmm0, [rsi]
                movsd   xmm1, [rsi+8]
                mov rax,2       ; 两个浮点数
                call printf
        leave
        ret
```

在此创建一个 dummy 变量来确保内存不是 16 字节对齐的。然后，在.data 段中使用 NASM 汇编器指令 align 16 以及在.bss 段中使用指令 alignb 16。你需要在每个需要对齐的数据块之前添加这些汇编器指令。

SSE 指令与未对齐的版本略有不同。我们使用 movaps(意思是"移动对齐打包单精度")将数据从内存复制到 xmm0。然后可以立即将内存中的打包数字添加到 xmm0 中的值。这与未对齐的版本不同，在未对齐的版本中，我们必须首先将这两个值放入一个 xmm 寄存器中。如果将 dummy 变量添加到未对齐的示例中，并尝试使用 movaps 而不是将内存变量作为第二个操作数的 movups，就可能出现运行时分段错误。

寄存器 xmm0 包含具有四个单精度值的结果和向量。然后，我们使用函数 printspfp 打印结果。在 printspfp 函数中，我们从内存中调用每个值，并将它们放入 xmm 寄存器中。因为 printf 需要双精度浮点参数，所以我们使用指令 cvtss2sd("将标量单精度转换为标量双精度")将单精度浮点数转换为双精度。

接下来，我们使用双精度值。这个过程类似于使用单精度，但是我们将 movapd 和 addpd 用于双精度值。

图 28-2 展示了对齐示例的输出。

```
jo@UbuntuDesktop:~/Desktop/linux64/gcc/34 sse_aligned$ make
nasm -f elf64 -g -F dwarf sse_aligned.asm -l sse_aligned.lst
gcc -o sse_aligned sse_aligned.o
jo@UbuntuDesktop:~/Desktop/linux64/gcc/34 sse_aligned$ ./sse_aligned
Single Precision Vector 1: 1.100000, 2.200000, 3.300000, 4.400000
Single Precision Vector 2: 1.100000, 2.200000, 3.300000, 4.400000
Sum of Single Precision Vector 1 and Vector 2: 2.200000, 4.400000, 6.600000, 8.800000
Double Precision Vector 1: 1.100000, 2.200000
Double Precision Vector 2: 3.300000, 4.400000
Sum of Double Precision Vector 1 and Vector 2: 4.400000, 6.600000
jo@UbuntuDesktop:~/Desktop/linux64/gcc/34 sse_aligned$
```

图 28-2　sse_aligned.asm 的输出

图 28-3 展示了未对齐的示例，其中添加了 dummy 变量作为 movaps 的第二个操作数。

```
jo@UbuntuDesktop:~/Desktop/linux64/gcc/34 sse_unaligned$ make
nasm -f elf64 -g -F dwarf sse_unaligned.asm -l sse_unaligned.lst
gcc -o sse_unaligned sse_unaligned.o -no-pie
jo@UbuntuDesktop:~/Desktop/linux64/gcc/34 sse_unaligned$ ./sse_unaligned
Single Precision Vector 1: 1.100000, 2.200000, 3.300000, 4.400000
Single Precision Vector 2: 1.100000, 2.200000, 3.300000, 4.400000
Segmentation fault (core dumped)
jo@UbuntuDesktop:~/Desktop/linux64/gcc/34 sse_unaligned$
```

图 28-3　sse_unaligned.asm 分段错误

28.3　小结

本章内容：
- 标量数据和打包数据
- 对齐和未对齐的数据
- 如何对齐数据
- 打包数据的数据移动和算术指令
- 如何在单精度和双精度数据之间进行转换

第 29 章

SSE 打包整数

在上一章中，我们使用了浮点值和指令。SSE 还提供了很多操作整数的指令，我们将在本章展示一些指令来帮助你入门。

29.1 适用于整数的 SSE 指令

代码清单 29-1 展示了一个示例程序。

代码清单 29-1：sse_integer.asm

```
; sse_integer.asm
extern printf

section .data
       dummy   db 13
align 16
       pdivector1   dd 1
                    dd 2
                    dd 3
                    dd 4
       pdivector2   dd 5
                    dd 6
                    dd 7
                    dd 8

fmt1 db "Packed Integer Vector 1: %d, %d, %d, %d",10,0
fmt2 db "Packed Integer Vector 2: %d, %d, %d, %d",10,0
fmt3 db "Sum Vector: %d, %d, %d, %d",10,0
```

```nasm
        fmt4 db "Reverse of Sum Vector: %d, %d, %d, %d",10,0
section .bss
alignb 16
        pdivector_res    resd 4
        pdivector_other  resd 4

section .text
        global main
main:
push rbp
mov rbp,rsp

; 打印向量 1
        mov rsi,pdivector1
        mov rdi,fmt1
        call printpdi
; 打印向量 2
        mov rsi,pdivector2
        mov rdi,fmt2
        call printpdi

; 添加两个对齐的双整数向量
        movdqa  xmm0, [pdivector1]
        paddd   xmm0, [pdivector2]

; 将结果保存在内存中
        movdqa  [pdivector_res], xmm0
; 打印内存中的向量
        mov rsi,pdivector_res
        mov rdi,fmt3
        call printpdi

; 将内存向量复制到 xmm3
        movdqa  xmm3,[pdivector_res]
; 从 xmm3 中提取打包值
        pextrd eax, xmm3, 0
        pextrd ebx, xmm3, 1
        pextrd ecx, xmm3, 2
        pextrd edx, xmm3, 3
; 以相反顺序插入 xmm0
        pinsrd xmm0, eax, 3
```

```
        pinsrd xmm0, ebx, 2
        pinsrd xmm0, ecx, 1
        pinsrd xmm0, edx, 0
; 打印反转后的向量
        movdqa [pdivector_other], xmm0
        mov rsi,pdivector_other
        mov rdi,fmt4
        call printpdi

; 退出
mov rsp,rbp
pop rbp
ret

;打印函数----------------------------------------
printpdi:
push rbp
mov rbp,rsp
        movdqa xmm0, [rsi]
        ; 从 xmm0 中提取打包值
            pextrd esi, xmm0,0
            pextrd edx, xmm0,1
            pextrd ecx, xmm0,2
            pextrd r8d, xmm0,3
        mov rax,0; 没有浮点数
        call printf
leave
ret
```

29.2 分析代码

这里有两个整数向量,我们使用 movdqa 指令将值复制到 xmm 寄存器中。该指令用于对齐的数据。然后 paddd 将寄存器中的值相加,并将结果放入 xmm0。要使用 printf,我们需要从 xmm 寄存器中提取整数值并将其放入"常规"寄存器中。请记住,在调用约定中,printf 会将 xmm 寄存器视为浮动寄存器。如果我们不提取整数值,printf 会将 xmm 寄存器中的值视为浮点值,并打印错误的值。对于打包整数的提取和插入,我们使用 pinsrd 和 pextrd。我们还反转了一个向量来展示如何将值插入 xmm 寄存器中的向量。

对于字节、字、双字和四字，movd、padd、pinsr 和 pextr 都有对应的版本。输出如图 29-1 所示。

```
jo@UbuntuDesktop:~/Desktop/linux64/gcc/35 sse_integer$ make
nasm -f elf64 -g -F dwarf sse_integer.asm -l sse_integer.lst
gcc -o sse_integer sse_integer.o
jo@UbuntuDesktop:~/Desktop/linux64/gcc/35 sse_integer$ ./sse_integer
Packed Integer Vector 1: 1, 2, 3, 4
Packed Integer Vector 2: 5, 6, 7, 8
Sum of Packed Integer Vector 1 and Vector 2: 6, 8, 10, 12
Reverse of Sum Vector: 12, 10, 8, 6
jo@UbuntuDesktop:~/Desktop/linux64/gcc/35 sse_integer$
```

图 29-1　sse_integer.asm 的输出

29.3　小结

本章内容：

- 整数打包数据
- 插入和提取打包整数的指令
- 复制打包整数并求和的指令

第 30 章

SSE 字符串操作

在 SSE 版本 4.2 中，引入了四个用于比较字符串的指令：两个用于隐式长度字符串的指令和两个用于显式长度字符串的指令。这四个指令中有两个使用掩码。

隐式长度的字符串是以 0 结尾的字符串。对于具有显式长度的字符串，必须通过其他方式指定长度。

我们将在本章花一些时间研究 SSE 字符串，因为比较指令较为复杂和特殊，尤其在使用掩码时。如表 30-1 所示。

表 30-1　SSE 字符串

字符串	指令	参数 1	参数 2	参数 3	输出
隐式	pcmpistri	xmm	xmm/m128	imm8	ecx 中的索引
隐式	pcmpistrm	xmm	xmm/m128	imm8	xmm0 中的掩码
显式	pcmpestri	xmm	xmm/m128	imm8	ecx 中的索引
显式	pcmpestrm	xmm	xmm/m128	imm8	xmm0 中的掩码

以下列出指令的含义。

- pcmpistri：打包比较隐式长度字符串，返回索引。
- pcmpistrm：打包比较隐式长度字符串，返回掩码。
- pcmpestri：打包比较显式长度字符串，返回索引。
- pcmpestrm：打包比较显式长度字符串，返回掩码。

这些比较指令有三个参数。参数 1 始终是 xmm 寄存器，参数 2 可以是 xmm 寄存器或内存位置，参数 3 是"即时数"，即指定指令执行方式的控制字节(在英特尔手册中为 imm8)。控制字节起着重要作用，因此我们将花一些时间来解释细节。

30.1 imm8 控制字节

表 30-2 展示了控制字节(Control Byte)的设计。

表 30-2 imm8 控制字节

选项	Bit 位置	Bit 值	运算	意义
	7	0	保留	保留
输出格式	6	0	位掩码	xmm0 包含 IntRes2 作为位掩码
		1	字节掩码	xmm0 包含 IntRes2 作为字节掩码
		0	最低有效索引	在 ecx 中发现的最低有效索引
		1	最高有效索引	在 ecx 中发现的最高有效索引
极性	5,4	00	+	IntRes2 = IntRes1
		01	-	IntRes2 = ~IntRes1
		10	掩码+	IntRes2 = IntRes1
		11	掩码-	IntRes2 = ~IntRes1
聚合和比较	3,2	00	等于任何	匹配字符
		01	范围内相等	匹配范围内的字符
		10	每个都相等	字符串比较
		11	顺序相等	子串搜索
数据格式	1,0	00	打包的无符号字节	
		01	打包的无符号双字	
		10	打包的有符号字节	
		11	打包的有符号双字	

比较指令接收输入数据(格式在位 1 和 0 中指定)，执行聚合和比较操作(位 2 和 3)，这将生成中间结果(arg1 和 arg2 之间的匹配)。该结果在英特尔手册中称为 IntRes1。将极性应用于 IntRes1 以获得 IntRes2。然后使用 IntRes2 以所需格式输出结果。负极性(~IntRes1)表示取 IntRes1 的补码并将结果放入 IntRes2。也就是说，将每个 1 位转换为 0 位，并将每个 0 位转换为 1 位。换言之，这是逻辑非(logical NOT)。IntRes2 中的结果可作为掩码存储在 xmm0 中，用于掩码指令 pcmpistrm 和 pcmpestrm，或作为索引存储在 ecx 中，用于 pcmpistri 和 pcmpestri。这里有一些例子会有所帮助。

以下是一些控制字节示例。

00001000 或 0x08:

  00——打包的无符号字节

10——每个都相等
00——正极性
00——ecx 中的最低有效索引

01000100 或 0x44：

00——打包的无符号字节
01——范围内相等
00——正极性
01——xmm0 包含字节掩码

30.2 使用 imm8 控制字节

30.2.1 位 0 和 1

位 0 和 1 表示数据源格式；数据源可以是打包的字节或打包的字(无符号或有符号)。

30.2.2 位 2 和 3

位 2 和 3 表示要应用的聚合。结果称为 IntRes1(中间结果 1)。从第二个操作数中提取一个 16 字节的块，并将其与第一个操作数中的内容进行比较。

聚合方式如下。

等于任意字符(00)或从一个集合中找到字符：这意味着搜索操作数 1 并在操作数 2 中查找任何字符。找到匹配项后，在 IntRes1 中将相应的位设置为 1。下面是一个例子。

```
操作数 1 : ""this is a joke!!"
操作数 2 : "i!"
IntRes1: 0010010000000011
```

范围内相等(01)或在一定范围内找到字符：这意味着搜索操作数 1 并查找操作数 2 中给定范围内的任何字符。找到匹配项后，在 IntRes1 中将相应的位设置为 1。下面是一个例子。

```
操作数 1 : "this is a joke!!"
操作数 2 : "aj"
```

217

IntRes1: 0010010010100100

每个都相等(10)或字符串比较：这意味着将操作数 1 中的任何字符与操作数 2 中的对应字符进行比较。找到匹配项后，将 IntRes1 中的对应位设置为 1。下面是一个例子。

```
操作数 1: "this is a joke!!"
操作数 2: "this is no joke!"
IntRes1: 1111111100000000
```

顺序相等(11)或子串搜索：这意味着在操作数 1 中搜索操作数 2 中的字符串。找到匹配项后，将 IntRes1 中的相应位设置为 1。下面是一个例子。

```
操作数 1: "this is a joke!!"
操作数 2: "is"
IntRes1: 0010010000000000
```

30.2.3 位 4 和 5

位 4 和 5 应用极性并将结果存储在 IntRes2 中。

正极(00)和(10)：IntRes2 将与 IntRes1 相同。下面是一个例子。

```
IntRes1: 0010010000000011
IntRes2: 0010010000000011
```

负极(01)和(11)：IntRes2 将是它们的补码，或者是 IntRes1 的逻辑取反。下面是一个例子。

```
IntRes1: 0010010000000011
IntRes2: 1101101111111100
```

30.2.4 位 6

位 6 设置输出格式，有两种情况。

(1) 不使用掩码
0：ecx 中返回的索引是 IntRes2 中设置的最低有效位。下面是一个例子。

```
IntRes2: 0010010011000000
ecx = 6
```

在 IntRes2 中，第一个 1 位位于索引 6 处（计数从 0 开始，从右边开始）。

1： ecx 中返回的索引是 IntRes2 中设置的最高有效位。下面是一个例子。

```
IntRes2: 0010010010100100
ecx = 13
```

在 IntRes2 中，最后一个 1 位位于索引 13 处（计数从 0 开始，从右边开始）。

(2) 使用掩码

IntRes2 作为 xmm0 的最低有效位中的一个掩码返回(零扩展至 128 位)。下面举一个例子。

```
在 string 字符串中搜索所有 'a' 和 'e' 字符
string = "qdacdekkfijlmdoz"
然后
xmm0: 024h
或者 0000000000100100(二进制形式)
```

请注意，掩码在 xmm0 中是相反的。

1： IntRes2 被扩展为 xmm0 的一个字节/字掩码。下面举一个例子。

```
在 string 字符串中搜索所有 'a' 和 'e' 字符
string = "qdacdekkfijlmdoz"
然后
xmm0: 00000000000000000000ff0000ff0000
```

请注意，掩码在 xmm0 中是相反的。

30.2.5 位 7

位 7 被保留。

30.2.6 标志

对于隐式长度指令，标志的使用方式与你在前面几章中看到的方式有所不同（请参阅英特尔手册）。

```
CF -如果 IntRes2 等于零则复位，否则设置
ZF -如果 xmm2/mem128 的任何字节/字为空，则设置，否则复位
```

SF -如果 xmm1 的任何字节/字为空，则设置，否则复位

OF - IntRes2[0]

AF -复位

PF -复位

对于显式长度指令，这些标志以不同方式使用，如下所示(请参阅英特尔手册)：

CF -如果 IntRes2 等于零则复位，否则设置

ZF -如果 EDX 的绝对值<16(8)，则设置；否则重置

SF -如果 EAX 的绝对值<16(8)，则设置；否则重置

OF - IntRes2 [0]

AF -复位

PF -复位

在下一章的示例中，我们将使用 CF 标志来查看是否存在任何结果，并使用 ZF 来检测字符串的结尾。

这个理论听起来可能很复杂。不必担心，很快我们就会做一些练习。

30.3 小结

本章内容：
- SSE 字符串操作指令
- imm8 控制字节的设计和使用

第 31 章

搜索字符

我们将在本章开始使用控制字节在字符串中查找特定字符。

31.1 确定字符串的长度

在第一个示例中，将通过查找终止 0 来确定字符串的长度。
代码清单 31-1 展示了示例代码。

代码清单 31-1：sse_string_length.asm

```
; sse_string_length.asm
extern printf
section .data
;模板           0123456789abcdef0123456789abcdef0123456789abcd e
;模板           12345678901234567890123456789012345678901234567
    string1 db "The quick brown fox jumps over the lazy river.",0
    fmt1 db  "This is our string: %s ",10,0
    fmt2 db  "Our string is %d characters long.",10,0
section .bss
section .text
    global main
main:
push  rbp
mov   rbp,rsp
    mov   rdi, fmt1
    mov   rsi, string1
    xor   rax,rax
```

```
        call printf
        mov  rdi, string1
        call pstrlen
        mov  rdi, fmt2
        mov  rsi, rax
        xor  rax,rax
        call printf
leave
ret
;用于计算字符串长度的函数------------------------
pstrlen:
push rbp
mov rbp,rsp
        mov   rax,   -16           ; 避免以后改变
        pxor xmm0, xmm0            ; 0 (字符串的终止)
.not_found:
        add        rax, 16         ; 避免在 pcmpistri 之后更改 ZF
        pcmpistri xmm0, [rdi + rax], 00001000b ;'equal each'
        jnz        .not_found      ; 是否找到 0?
        add        rax, rcx        ; rcx 包含 0 的索引
        inc        rax             ; 更正为从索引 0 开始
leave
ret
```

在程序开始时，我们在注释中添加了两个模板，以使字符计数更容易。一个模板使用十进制编号，从 1 开始；另一个模板使用十六进制编号，从索引 0 开始。

```
;template    12345678901234567890123456789012345678901234567
;template    0123456789abcdef0123456789abcdef0123456789abcd e
string1 db  "The quick brown fox jumps over the lazy river.",0
```

首先，我们像往常一样打印字符串。然后调用定制的搜索函数 pstrlen。该函数扫描第一次出现的零字节。指令 pcmpistri 一次分析 16 字节的块；我们使用 rax 作为块计数器。如果 pcmpistri 在当前块中检测到一个零字节，将设置 ZF 并决定是否跳转。在评估跳转之前增加 rax 会影响 ZF 标志；为避免这种情况，必须在 pcmpistri 之前增加 ZF 标志。这就是为什么在 rax 中以-16 开始；现在可在使用 pcmpistri 之前增加 rax。注意 pxor 指令，它是 xmm 寄存器的"逻辑或"指令。SIMD 有自己的逻辑指令！

瞬时控制字节包含 00001000，这意味着：

- 00 打包的无符号字节
- 10 每个都相等
- 00 正极性
- 0 最低有效指数
- 0 保留

你可能希望我们使用 equal any 来找到任何 0，但我们使用的是 equal each！这是为什么？

你必须知道 pcmpistri 会将 rcx 的值初始化为 16，该值是一个块中的字节数。如果找到匹配的字节，则 pcmpistri 将在 rcx 中复制匹配字节的索引。如果找不到匹配项，则 rcx 中的值将保持为 16。

请参阅英特尔手册，特别是卷 2B。第 4.1.6 节解释了当块具有"无效"字节或超过字符串结尾的字节时会发生什么。

可以使用表 31-1 进行解释。

表 31-1 处理方式

xmm0	内存	equal any	equal each
无效	无效	强制假	强制真
无效	有效	强制假	强制假

xmm0 之所以无效，因为我们将它初始化为包含 0 字节。当我们有一个包含 0 字节的 16 字节块时，在 equal any 的情况下，pcmpistri 检测到 16 字节中的一个包含 0。此时，xmm0 和内存都无效。但是，pcmpistri 被设计成在 equal any 的情况下"强制假"。因此，pcmpistri 认为不存在匹配，并在 rcx 中返回 16，因此计算的字符串长度将不正确。

当我们使用 equal each 时，xmm0 像以前一样是无效的，并且只要 pcmpistri 读取了块中的终止 0 字节，它就会被设计为"强制真"。0 字节的索引被记录在 ecx 中。ecx 中的值可用于正确计算字符串的结尾。

注意，程序读取 16 个字节的块，只要找到数据的位置在分配给程序的内存空间内即可。如果它试图读取的范围超出允许的内存边界，程序将崩溃。你可以通过跟踪内存页面中的位置来避免这种情况(大多数情况下，页面是 4K 字节的块)，如果你接近页面边界，则开始按字节读取字节。这样，你将永远不会意外地尝试从允许的内存页面切换到另一个进程的内存页面。我们没有实现这个功能，以免使说明和示例程序变得复杂。但请注意，这种情况可能会发生。

输出如图 31-1 所示。如你所见，字符串长度包括终止 null。

```
jo@UbuntuDesktop:~/Desktop/linux64/gcc/36 sse_string_length$ make
nasm -f elf64 -g -F dwarf sse_string_length.asm -l sse_string_length.lst
gcc -o sse_string_length sse_string_length.o -no-pie
jo@UbuntuDesktop:~/Desktop/linux64/gcc/36 sse_string_length$ ./sse_string_length
This is our string: The quick brown fox jumps over the lazy river.
Our string is 47 characters long.
jo@UbuntuDesktop:~/Desktop/linux64/gcc/36 sse_string_length$
```

图 31-1　sse_string_search.asm 的输出结果

31.2　在字符串中搜索

现在我们知道了如何确定字符串的长度，下面在字符串中进行一些搜索(参见代码清单 31-2)。

代码清单 31-2：sse_string_search.asm

```
; sse_string_search.asm
extern printf
section .data
;模板          12345678901234567890123456789012345  6
;模板          0123456789abcdef0123456789abcdef0123456789abc  d
string1 db "the quick brown fox jumps over the lazy river",0
string2 db "e",0
fmt1    db "This is our string: %s ",10,0
fmt2    db "The first '%s' is at position %d.",10,0
fmt3    db "The last '%s' is at position %d.",10,0
fmt4    db "The character '%s' didn't show up!.",10,0
section .bss
section .text
    global main
main:
push rbp
mov rbp,rsp
    mov rdi, fmt1
    mov rsi, string1
    xor rax,rax
    call printf

;找到第一次出现
    mov rdi, string1
    mov rsi, string2
```

```
        call pstrscan_f
        cmp  rax,0
        je   no_show
        mov  rdi, fmt2
        mov  rsi, string2
        mov  rdx, rax
        xor  rax,rax
        call printf
; 找到最后一次出现
        mov  rdi, string1
        mov  rsi, string2
        call pstrscan_l
        mov  rdi, fmt3
        mov  rsi, string2
        mov  rdx, rax
        xor  rax,rax
        call printf
        jmp  exit
no_show:
        mov  rdi, fmt4
        mov  rsi, string2
        xor  rax, rax
        call printf
exit:
leave
ret
;------ 找到第一次出现----------------------
pstrscan_f:
push rbp
mov rbp,rsp
        xor    rax, rax
        pxor   xmm0, xmm0
        pinsrb xmm0, [rsi],0
.block_loop:
        pcmpistri xmm0, [rdi + rax], 00000000b
        jc   .found
        jz   .none
        add  rax, 16
```

```
        jmp     .block_loop
.found:
        add     rax, rcx        ; rcx 包含字符的位置
        inc     rax             ; 从1(而不是0)开始计数
leave
ret
.none:
        xor     rax,rax         ; 没有找到，返回0
leave
ret
;------ 找到最后一次出现----------------------
pstrscan_l:
push rbp
mov rbp,rsp
push rbx            ; 被调用者保留
push r12            ; 被调用者保留
        xor     rax, rax
        pxor    xmm0, xmm0
        pinsrb  xmm0, [rsi],0
        xor     r12,r12
.block_loop:
        pcmpistri xmm0, [rdi + rax], 01000000b
        setz    bl
        jc      .found
        jz      .done
        add     rax, 16
        jmp     .block_loop
.found:
        mov     r12, rax
        add     r12, rcx        ; rcx 包含字符的位置
        inc     r12
        cmp     bl,1
        je      .done
        add     rax,16
        jmp     .block_loop
pop r12             ; 被调用者保留
pop rbx             ; 被调用者保留
leave
ret
```

```
.done:
        mov rax,r12
pop r12             ; 被调用者保留
pop rbx             ; 被调用者保留
leave
ret
```

在程序开始时，我们在注释中添加了两个模板，以使字符计数更容易。

这里，string1 包含字符串，string2 包含搜索参数。我们将搜索参数的第一个和最后一个匹配项。首先打印字符串，然后调用自定义函数。我们有单独的函数来查找字符的第一次出现和最后一次出现。函数 pstrscan_f 查找搜索参数的第一次出现。pcmpistri 指令一次处理 16 个字节的块；我们使用 rax 作为块计数器，使用 pxor 指令清除 xmm0。我们使用 pinsrb 将搜索参数放在 xmm0 的低位字节(字节 0)。我们使用 equal any 来查找出现的位置，一旦找到出现的位置，rcx 表示当前 16 字节块中匹配字节的索引。如果在当前块中未找到匹配项，则将值 16 放入 rcx 中。我们使用 jc 检查 CF 是否为 1。如果是 1，表示我们找到了一个匹配项。rcx 被添加到 rax，其中包含先前块中已经筛选的字节数，然后返回 rax，将其更正为计数从 1(而不是 0)开始。

如果 CF=0，使用 jz 检查是否到达最后一个块。当检测到空字节时，pcmpistri 会将 ZF 设置为 1，并清除 rax，因为找不到匹配项。函数返回 0。

当然，我们没有进行任何错误检查；如果字符串不是以 null 终止的，则可能得到错误结果。尝试删除字符串末尾的 0 并观察结果。

函数 pstrscan_l 查找搜索参数的最后一个匹配项。这比仅寻找第一个匹配并退出更复杂。我们必须读取所有 16 字节的块，并跟踪在块中的最后一次出现。因此，即使找到了一次出现，也必须继续循环，直至找到终止 0 为止。为了关注终止 0，在检测到 0 后将寄存器 bl 设置为 1。寄存器 r12 用于记录最近匹配的索引。请参见图 31-2。

```
jo@UbuntuDesktop:~/Desktop/linux64/gcc/36 sse_string_search$ make
nasm -f elf64 -g -F dwarf sse_string_search.asm -l sse_string_search.lst
gcc -o sse_string_search sse_string_search.o -no-pie
jo@UbuntuDesktop:~/Desktop/linux64/gcc/36 sse_string_search$ ./sse_string_search
This is our string: the quick brown fox jumps over the lazy river
The first 'e' is at position 3.
The last 'e' is at position 44.
jo@UbuntuDesktop:~/Desktop/linux64/gcc/36 sse_string_search$
```

图 31-2　sse_string_search.asm 的输出

31.3 小结

本章内容：
- 使用 pcmpistri 查找字符和字符串长度
- 用不同的控制字节解释 pcmpistri 的结果

第 32 章

比较字符串

在上一章中，我们使用了具有隐式长度的字符串，这意味着这些字符串以空字节终止。在本章中，将比较具有隐式长度的字符串和具有显式长度的字符串。

32.1 隐式长度

我们将查找不同的字符，而不是匹配字符。代码清单 32-1 显示了要讨论的示例代码。

代码清单 32-1：sse_string2_imp.asm

```
; sse_string2_imp.asm
; 比较字符串的隐式长度
extern printf
section .data
    string1    db  "the quick brown fox jumps over the lazy"
               db  " river",10,0
    string2    db  "the quick brown fox jumps over the lazy"
               db  " river",10,0
    string3    db  "the quick brown fox jumps over the lazy"
               db  " dog",10,0
    fmt1   db "Strings 1 and 2 are equal.",10,0
    fmt11  db "Strings 1 and 2 differ at position %i.",10,0
    fmt2   db "Strings 2 and 3 are equal.",10,0
    fmt22  db "Strings 2 and 3 differ at position %i.",10,0
section .bss
section .text
```

```
        global main
main:
push rbp
mov rbp,rsp
;首先打印字符串
        mov   rdi, string1
        xor   rax,rax
        call  printf
        mov   rdi, string2
        xor   rax,rax
        call  printf
        mov   rdi, string3
        xor   rax,rax
        call  printf
;比较 string1 和 string2
        mov   rdi, string1
        mov   rsi, string2
        call  pstrcmp
        mov   rdi,fmt1
        cmp   rax,0
        je    eql1              ;两个字符串相同
        mov   rdi,fmt11         ;两个字符串不相同
eql1:
        mov   rsi, rax
        xor   rax,rax
        call  printf
;比较 string2 和 string3
        mov   rdi, string2
        mov   rsi, string3
        call  pstrcmp
        mov   rdi,fmt2
        cmp   rax,0
        je    eql2              ;两个字符串相同
        mov   rdi,fmt22         ;两个字符串不相同
eql2:
        mov rsi, rax
        xor   rax,rax
        call  printf
;退出
```

```
        leave
        ret
        ;字符串比较-------------------------------------------
pstrcmp:
push rbp
mov rbp,rsp
        xor     rax, rax              ;
        xor     rbx, rbx              ;
.loop:  movdqu      xmm1, [rdi + rbx]
        pcmpistri   xmm1, [rsi + rbx], 0x18  ; equal each | neg.polarity
        jc          .differ
        jz          .equal
        add         rbx, 16
        jmp         .loop
.differ:
        mov rax,rbx
        add rax,rcx        ;不同字符的位置
        inc  rax           ;因为索引从 0 开始
.equal:
        leave
        ret
```

像往常一样,首先打印字符串;然后调用函数 pstrcmp 来比较字符串。基本信息在 pstrcmp 函数中。控制字节是 0x18 或 00011000,即从右到左:打包的整数字节、每个字节相等(equal each)、负极性以及 ecx(包含第一次出现的索引)。pcmpistri 指令使用如下这些标志(你可以在英特尔手册中找到这些内容)。

CFlag:如果 IntRes2 等于零,则复位;否则设置。

ZFFLAG:如果 xmm2/mem128 的任何字节/字为空则设置;否则复位。

SFlag:如果 xmm1 的任何字节/字为空,则设置;否则复位。

OFlag:IntRes2[0]。

AFlag:复位。

PFlag:复位。

在此示例中,pcmpistri 为每个匹配在 IntRes1 中的相应位置放入一个 1。当找到一个不同的字节时,会将 0 写入 IntRes1 中的相应位置。然后形成 IntRes2,并将负极性应用于 IntRes1。IntRes2 将在不同的索引(负极性)处包含 1,因此 IntRes2 不会为零,并且 CF 将设置为 1。然后,循环将被中断,并且 pstrcmp 将返回 rax 中不同字符的位置。如果未设置 CF,但 pcmpistri 检测到终止零,则函数将在 rax 返回 0。

图 32-1 展示了输出。

```
jo@ubuntu18:~/Desktop/Book/37 sse_string2_imp$ ./sse_string2
the quick brown fox jumps over the lazy river
the quick brown fox jumps over the lazy river
the quick brown fox jumps over the lazy dog
Strings 1 and 2 are equal.
Strings 2 and 3 differ at position 41.
jo@ubuntu18:~/Desktop/Book/37 sse_string2_imp$
```

图 32-1　sse_string2_imp.asm 的输出

32.2　显式长度

大多数情况下，我们使用隐式长度的字符串，但代码清单 32-2 展示了一个显式长度字符串的示例。

代码清单 32-2：sse_string3_exp.asm

```
; sse_string3_exp.asm
; 比较字符串显式长度
extern printf
section .data
    string1    db  "the quick brown fox jumps over the "
               db  "lazy river"
    string1Len equ $ - string1
    string2    db  "the quick brown fox jumps over the "
               db  "lazy river"
    string2Len equ $ - string2
    dummy      db "confuse the world"
    string3    db  "the quick brown fox jumps over the "
               db  "lazy dog"
    string3Len equ $ - string3

    fmt1       db  "Strings 1 and 2 are equal.",10,0
    fmt11      db  "Strings 1 and 2 differ at position %i.",10,0
    fmt2       db  "Strings 2 and 3 are equal.",10,0
    fmt22      db  "Strings 2 and 3 differ at position %i.",10,0

section .bss
    buffer resb 64
section .text
    global main
main:
```

```
        push    rbp
        mov     rbp,rsp
        ; 比较 string1 和 string2
            mov     rdi, string1
            mov     rsi, string2
            mov     rdx, string1Len
            mov     rcx, string2Len
            call    pstrcmp
            push    rax             ;将结果压入堆栈供以后使用

; 打印 string1、string2 以及结果
;-----------------------------------------------------------
; 首先用换行符构建字符串并以 0 终止
; string1
            mov     rsi,string1
            mov     rdi,buffer
            mov     rcx,string1Len
            rep     movsb
            mov     byte[rdi],10    ; 将 NL 添加到 buffer
            inc     rdi             ; 将终止 0 添加到 buffer
            mov     byte[rdi],0
;打印
            mov     rdi, buffer
            xor     rax,rax
            call    printf
; string2
            mov     rsi,string2
            mov     rdi,buffer
            mov     rcx,string2Len
            rep     movsb
            mov     byte[rdi],10    ; 将 NL 添加到 buffer
            inc     rdi             ; 将终止 0 添加到 buffer
            mov     byte[rdi],0
;打印
            mov     rdi, buffer
            xor     rax,rax
            call    printf
;-----------------------------------------------------------
; 打印比较结果
```

```
        pop   rax           ;调用返回值
        mov   rdi,fmt1
        cmp   rax,0
        je    eql1
        mov   rdi,fmt11
eql1:
        mov   rsi, rax
        xor   rax,rax
        call  printf
;------------------------------------------------------------
;------------------------------------------------------------
; 比较 string2 和 string3
        mov   rdi, string2
        mov   rsi, string3
        mov   rdx, string2Len
        mov   rcx, string3Len
        call  pstrcmp
        push  rax
; 打印 string3 和结果
;------------------------------------------------------------
; 首先用换行符构建字符串并以 0 终止
; string3
        mov   rsi,string3
        mov   rdi,buffer
        mov   rcx,string3Len
        rep   movsb
        mov   byte[rdi],10      ; 将 NL 添加到 buffer
        inc   rdi               ; 将终止 0 添加到 buffer
        mov   byte[rdi],0
;打印
        mov   rdi, buffer
        xor   rax,rax
        call  printf
;------------------------------------------------------------
; 打印比较结果
        pop   rax           ;调用返回值
        mov   rdi,fmt2
        cmp   rax,0
        je    eql2
```

```
            mov     rdi,fmt22
    eql2:
            mov     rsi, rax
            xor     rax,rax
            call    printf

    ; 退出
    leave
    ret
    ;-----------------------------------------------------------
    pstrcmp:
    push rbp
    mov rbp,rsp
            xor     rbx, rbx
            mov     rax,rdx         ;rax 包含第一个字符串的长度
            mov     rdx,rcx         ;rdx 包含第二字符串的长度
            xor     rcx,rcx         ;rcx 作为索引
    .loop:
            movdqu   xmm1, [rdi + rbx]
            pcmpestri xmm1, [rsi + rbx], 0x18  ; equal each|neg.polarity
            jc      .differ
            jz      .equal
            add     rbx, 16
            sub     rax,16
            sub     rdx,16
            jmp     .loop
    .differ:
            mov     rax,rbx
            add     rax,rcx         ; rcx 包含不同的位置
            inc     rax             ; 计数器从 0 开始
            jmp     exit
    .equal:
            xor     rax,rax
    exit:
    leave
    ret
```

如你所见，使用显式长度有时会使事情复杂化。那为什么还要使用它呢？许多通信协议都使用它，或者你的应用程序可能要求在数据中使用 0。我们必须以另一

种方式提供字符串的长度。在本例中，我们从 .data 段中的内存位置计算字符串的长度。但是，printf 需要以零结尾的字符串。因此，在我们演示了如何比较具有显式长度的字符串之后，将在缓冲区中重建字符串，在缓冲区中添加一个换行符和一个终止 null，然后将缓冲区移交给 printf。

现在看一下比较函数 pstrcmp。第一个字符串的长度进入 rax，第二个字符串的长度进入 rdx。然后开始一个循环： 将 16 字节块的地址加载到 xmm1 寄存器中，并调用 pcmpestri，控制字节为 0x18。接下来，下面看看标志，你可以在英特尔手册中找到以下内容。

CFlag：如果 IntRes2 等于零，则复位；否则设置。

ZFlag：如果 EDX 的绝对值小于 16(8)，则设置；否则复位。

SFlag：如果 EAX 的绝对值小于 16(8)，则设置；否则复位。

OFlag：IntRes2 [0]。

AFlag：复位。

PFlag：复位。

请注意，pcmpestri 和 pcmpistri 使用 ZF 和 SF 的方式不同。不同于 ZF 发出一个终止 null 信号，在每个循环中，我们将减少 rax 和 rdx，当其中一个低于 16 时，循环终止。

图 32-2 展示了输出。

```
jo@ubuntu18:~/Desktop/Book/38_0_sse_string3_exp$ ./sse_string3
the quick brown fox jumps over the lazy river
the quick brown fox jumps over the lazy river
Strings 1 and 2 are equal.
the quick brown fox jumps over the lazy dog
Strings 2 and 3 differ at position 41.
jo@ubuntu18:~/Desktop/Book/38_0_sse_string3_exp$
```

图 32-2　sse_string3_exp.asm 的输出

32.3　小结

本章内容：

- 隐式和显式字符串长度
- 负极性
- 使用标志

第 33 章

重排

对于无掩码字符串指令,我们有几个选项。我们可以找到一个字符的第一次或最后一次出现,但找到所有出现的情况更具挑战性。我们可以比较字符串并找出差异,但是要找到所有差异则更复杂。幸运的是,我们还有使用掩码的字符串指令,这使它们更加强大。但是,在深入了解掩码指令之前,我们需要研究一下重排(shuffle)。

33.1 重排初探

重排是指对打包的值进行移动操作。移动可以在同一 xmm 寄存器内,可以从一个 xmm 寄存器到另一个 xmm 寄存器,也可以从一个 128 位内存位置到一个 xmm 寄存器。

代码清单 33-1 显示了示例代码。

代码清单 33-1:shuffle.asm

```
; shuffle.asm
extern printf
section .data
    fmt0    db "These are the numbers in memory: ",10,0
    fmt00   db "This is xmm0: ",10,0
    fmt1    db "%d ",0
    fmt2    db "Shuffle-broadcast double word %i:",10,0
    fmt3    db "%d %d %d %d",10,0
    fmt4    db "Shuffle-reverse double words:",10,0
    fmt5    db "Shuffle-reverse packed bytes in xmm0:",10,0
```

```nasm
        fmt6    db "Shuffle-rotate left:",10,0
        fmt7    db "Shuffle-rotate right:",10,0
        fmt8    db "%c%c%c%c%c%c%c%c%c%c%c%c%c%c%c%c",10,0
        fmt9    db "Packed bytes in xmm0:",10,0
        NL      db 10,0

        number1 dd 1
        number2 dd 2
        number3 dd 3
        number4 dd 4

        char db "abcdefghijklmnop"
        bytereverse db 15,14,13,12,11,10,9,8,7,6,5,4,3,2,1,0

section .bss
section .text
        global main
main:
push rbp
mov rbp,rsp
        sub rsp,32          ;原始 xmm0 和修改后的 xmm0 的堆栈空间
; 重排双字(DOUBLE WORDS)
; 首先反向打印数字
        mov     rdi, fmt0
        call    printf
        mov     rdi, fmt1
        mov     rsi, [number4]
        xor     rax,rax
        call    printf
        mov     rdi, fmt1
        mov     rsi, [number3]
        xor     rax,rax
        call    printf
        mov     rdi, fmt1
        mov     rsi, [number2]
        xor     rax,rax
        call    printf
        mov     rdi, fmt1
        mov     rsi, [number1]
        xor     rax,rax
```

```
        call    printf
        mov     rdi, NL
        call    printf

; 用数字构建 xmm0
        pxor    xmm0,xmm0
        pinsrd  xmm0, dword[number1],0
        pinsrd  xmm0, dword[number2],1
        pinsrd  xmm0, dword[number3],2
        pinsrd  xmm0, dword[number4],3
        movdqu  [rbp-16],xmm0           ;保存 xmm0 以备后用
        mov     rdi, fmt00
        call    printf          ;print title
        movdqu  xmm0,[rbp-16]           ;printf 之后恢复 xmm0
        call    print_xmm0d             ;打印 xmm0
        movdqu  xmm0,[rbp-16]           ;printf 之后恢复 xmm0

; SHUFFLE-BROADCAST(重排-广播)
; 重排：广播最低有效双字(索引 0)
        movdqu  xmm0,[rbp-16]           ;恢复 xmm0
        pshufd  xmm0,xmm0,00000000b  ;重排
        mov     rdi,fmt2
        mov     rsi, 0                  ;打印标题
        movdqu  [rbp-32],xmm0           ;printf 销毁 xmm0
        call    printf
        movdqu  xmm0,[rbp-32]           ;在 printf 之后恢复 xmm0
        call    print_xmm0d             ;打印 xmm0 的内容

; 重排：广播双字(索引 1)
        movdqu  xmm0,[rbp-16]           ;恢复 xmm0
        pshufd  xmm0,xmm0,01010101b  ;重复
        mov     rdi,fmt2
        mov     rsi, 1                  ;打印标题
        movdqu  [rbp-32],xmm0           ;printf 销毁 xmm0
        call    printf
        movdqu  xmm0,[rbp-32]           ;在 printf 之后恢复 xmm0
        call    print_xmm0d             ;打印 xmm0 的内容

; 重排：广播双字(索引 2)
        movdqu  xmm0,[rbp-16]           ;恢复 xmm0
        pshufd  xmm0,xmm0,10101010b  ;重排
```

```asm
        mov     rdi,fmt2
        mov     rsi, 2              ;打印标题
        movdqu  [rbp-32],xmm0       ;printf 销毁 xmm0
        call    printf
        movdqu  xmm0,[rbp-32]       ;在 printf 之后恢复 xmm0
        call    print_xmm0d         ;打印 xmm0 的内容

; 重排：广播双字(索引 3)
        movdqu  xmm0,[rbp-16]       ;恢复 xmm0
        pshufd  xmm0,xmm0,11111111b ;重排
        mov     rdi,fmt2
        mov     rsi, 3              ;打印标题
        movdqu  [rbp-32],xmm0       ;printf 销毁 xmm0
        call    printf
        movdqu  xmm0,[rbp-32]       ;在 printf 之后恢复 xmm0
        call    print_xmm0d         ;打印 xmm0 的内容

; SHUFFLE-REVERSE(重排-反向)
; 反转双字
        movdqu  xmm0,[rbp-16]       ;恢复 xmm0
        pshufd  xmm0,xmm0,00011011b ;重排
        mov     rdi,fmt4
        movdqu  [rbp-32],xmm0       ;打印标题
                                    ;printf 销毁 xmm0
        call    printf
        movdqu  xmm0,[rbp-32]       ;在 printf 之后恢复 xmm0
        call    print_xmm0d         ;打印 xmm0 的内容

; SHUFFLE-ROTATE(重排-旋转)
; 向左旋转
        movdqu  xmm0,[rbp-16]       ;恢复 xmm0
        pshufd  xmm0,xmm0,10010011b ;重排
        mov     rdi,fmt6            ;打印标题
        movdqu  [rbp-32],xmm0       ;printf 销毁 xmm0
        call    printf
        movdqu  xmm0,[rbp-32]       ;在 printf 之后恢复 xmm0
        call    print_xmm0d         ;打印 xmm0 的内容

; 向右旋转
        movdqu  xmm0,[rbp-16]       ;恢复 xmm0
        pshufd  xmm0,xmm0,00111001b ;重排
        mov     rdi,fmt7            ;打印标题
```

第 33 章 ■ 重排

```
        movdqu  [rbp-32],xmm0       ;printf 销毁 xmm0
        call    printf
        movdqu  xmm0,[rbp-32]       ;在 printf 之后恢复 xmm0
        call    print_xmm0d         ;打印 xmm0 的内容
;SHUFFLING BYTES(重排字节)
        mov     rdi, fmt9
        call    printf              ;打印标题
        movdqu  xmm0,[char]         ;在 xmm0 中加载字符
        movdqu  [rbp-32],xmm0       ;printf 销毁 xmm0
        call    print_xmm0b         ;打印 xmm0 的字节
        movdqu  xmm0,[rbp-32]       ;在 printf 之后恢复 xmm0
        movdqu  xmm1,[bytereverse]  ;加载掩码
        pshufb  xmm0,xmm1           ;重排字节
        mov     rdi,fmt5            ;打印标题
        movdqu  [rbp-32],xmm0       ;printf 销毁 xmm0
        call    printf
        movdqu  xmm0,[rbp-32]       ;在 printf 之后恢复 xmm0
        call    print_xmm0b         ;打印 xmm0 的内容
leave
ret
;用于打印双字的函数--------------------
print_xmm0d:
push rbp
mov  rbp,rsp
        mov     rdi, fmt3
        xor     rax,rax
        pextrd  esi, xmm0,3         ;反向提取双字，小端格式
        pextrd  edx, xmm0,2
        pextrd  ecx, xmm0,1
        pextrd  r8d, xmm0,0
        call    printf
leave
ret
;用于打印字节的函数--------------------------
print_xmm0b:
push rbp
mov  rbp,rsp
        mov     rdi, fmt8
```

```
        xor     rax,rax
        pextrb  esi, xmm0,0       ;反向，小端格式
        pextrb  edx, xmm0,1       ;首先使用寄存器
        pextrb  ecx, xmm0,2       ;然后是堆栈
        pextrb  r8d, xmm0,3
        pextrb  r9d, xmm0,4
        pextrb  eax, xmm0,15
        push    rax
        pextrb  eax, xmm0,14
        push    rax
        pextrb  eax, xmm0,13
        push    rax
        pextrb  eax, xmm0,12
        push    rax
        pextrb  eax, xmm0,11
        push    rax
        pextrb  eax, xmm0,10
        push    rax
        pextrb  eax, xmm0,9
        push    rax
        pextrb  eax, xmm0,8
        push    rax
        pextrb  eax, xmm0,7
        push    rax
        pextrb  eax, xmm0,6
        push    rax
        pextrb  eax, xmm0,5
        push    rax
        xor     rax,rax
        call    printf
    leave
    ret
```

首先，我们在堆栈上为 128 个字节的变量保留空间。我们需要这个空间将 xmm 寄存器压入堆栈。我们不能在 xmm 寄存器中使用标准的压入/弹出(push/pop)指令；必须使用内存寻址将它们复制到堆栈或从堆栈复制。我们使用基指针 rbp 作为参考点。

我们打印将用作打包值的数字。然后使用 pinsrd(意思是"打包插入双精度")指令将数字作为双字加载到 xmm0 中。使用指令 movdqu[rbp-16], xmm0 将 xmm0 保

存(压入)为一个局部堆栈变量(我们在程序开始时为这个局部变量保留了空间)。每次执行 printf 时，xmm0 都会被修改，不管是有意还是无意。因此，如有必要，我们必须保留和恢复 xmm0 的原始值。指令 movdqu 用于移动未对齐的打包整数值。为了帮助可视随机化的结果，我们在打印时使用了 little-endian(小端)格式。这样做将显示 xmm0，正如你在 SASM 之类的调试器中所看到的。要进行重排，我们需要一个目标操作数、一个源操作数和一个重排掩码。掩码是 8 位即时数。在以下各节中，我们将讨论一些有用的重排示例以及相应的掩码。

- 重排广播
- 重排反转
- 重排旋转

33.2 重排广播

图解可以让一切变得更容易理解。图 33-1 显示了四个重排广播的例子。

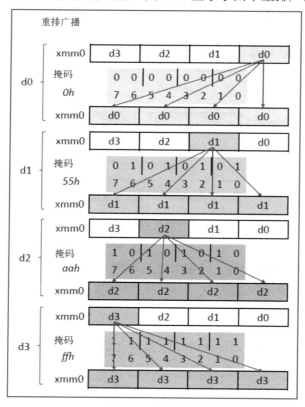

图 33-1 重排广播

上图中，源和目标都是 xmm0。最低有效双字 d0 在掩码中指定为 00b。第二低有效双字 d1 指定为 01b。第三低有效双字 d2 指定为 10b。第四低有效双字 d3 指定为 11b。二进制掩码 10101010b(或十六进制的 aah)的工作方式如下：将 d2(10b)放在四个目标打包双字位置。类似地，掩码 11111b 将 d3(11b)放置在四个目标打包双字位置。

当你研究代码时，将看到以下简单的重排指令：

```
pshufd xmm0,xmm0,10101010b
```

我们完成了 xmm0 中第三低元素的广播。因为 printf 函数修改了 xmm0，所以需要在调用 printf 之前通过将 xmm0 的内容存储到内存来保存它。事实上，需要做更多工作来保护 xmm0 的内容，而不是重排本身。

当然，并不局限于这里介绍的四个掩码；你可以创建任何 8 位掩码，并根据需要进行混合和重排。

33.3 重排反转

图 33-2 展示了重排反转的示意图。

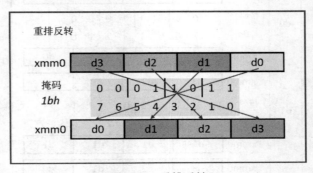

图 33-2　重排反转

掩码为 00011011b 或 1bh，转换为以下内容。

- 11(d3 中的值)进入位置 0
- 01(d2 中的值)进入位置 1
- 10(d1 中的值)进入位置 2
- 00(d0 中的值)进入位置 3

正如你在示例代码中看到的，用汇编语言编写代码很简单，如下所示：

```
pshufd xmm0,xmm0,1bh
```

33.4 重排旋转

重排旋转有两个版本：左旋和右旋。只需要提供正确的掩码作为重排指令的最后一个参数即可。原理概述如图33-3所示。

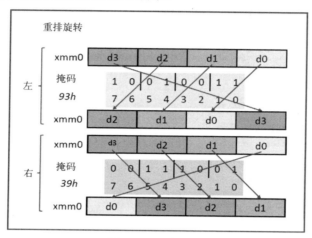

图 33-3 重排旋转

汇编语言代码如下：

```
pshufd xmm0, xmm0, 93h
pshufd xmm0, xmm0, 39h
```

33.5 重排字节

你可以使用 pshufd 和 pshufw 重排双字。也可分别用 pshufhw 和 pshuflw 重排高位单字和低位单字。你可以在英特尔手册中找到所有详细信息。所有这些指令都使用源操作数、目标操作数和用即时数指定的掩码。提供即时数作为掩码有其局限性：它不灵活，你必须在汇编时(而不是在运行时)提供掩码。

但还是有一个解决方案：对字节进行重排。

可以使用 pshufb 重排字节。该指令仅使用两个操作数：目标 xmm 寄存器操作数和存储在 xmm 寄存器或128位内存位置中的掩码。在前面的代码中，我们用 pshufb 反转了字符串'char'。我们在.data 段的 bytereverse 内存位置提供了一个掩码；掩码要求我们将字节15 放在位置 0，将字节14 放在位置1，以此类推。我们在 xmm0 中复制要重排的字符串，在 xmm1 中复制掩码，因此重排指令如下：

```
pshufb xmm0, xmm1
```

然后奇迹就发生了。请记住，掩码放在第二个操作数中；源与目标相同，放在第一个操作数中。

这样的好处是，我们不必在汇编时立即提供掩码。作为运行时计算的结果，可以在 xmm1 中内置掩码。

最后，图 33-4 展示了示例代码的输出。

```
jo@UbuntuDesktop:~/Desktop/linux64/gcc/38_1 shuffle$ make
nasm -f elf64 -g -F dwarf shuffle.asm -l shuffle.lst
gcc -o shuffle shuffle.o -no-pie
jo@UbuntuDesktop:~/Desktop/linux64/gcc/38_1 shuffle$ ./shuffle
These are the numbers in memory:
4 3 2 1
This is xmm0:
4 3 2 1
Shuffle-broadcast double word 0:
1 1 1 1
Shuffle-broadcast double word 1:
2 2 2 2
Shuffle-broadcast double word 2:
3 3 3 3
Shuffle-broadcast double word 3:
4 4 4 4
Shuffle-reverse double words:
1 2 3 4
Shuffle-rotate left:
3 2 1 4
Shuffle-rotate right:
1 4 3 2
Packed bytes in xmm0:
abcdefghijklmnop
Shuffle-reverse packed bytes in xmm0:
ponmlkjihgfedcba
jo@UbuntuDesktop:~/Desktop/linux64/gcc/38_1 shuffle$
```

图 33-4　shuffle.asm 的输出

33.6　小结

本章内容：

- 重排指令
- 重排掩码
- 运行时掩码
- 如何将堆栈与 xmm 寄存器一起使用

第 34 章

SSE 字符串掩码

现在我们知道如何重排了，下面开始学习字符串掩码。

请记住，SSE 提供了两个使用掩码的字符串操作指令：pcmpistrm 和 pcmpestrm。我们将使用隐式长度指令。刚开始使用掩码看起来很复杂，但是一旦掌握了窍门，就会发现掩码的功能是多么强大。

34.1 搜索字符

代码清单 34-1、代码清单 34-4 和代码清单 34-3 展示了这个示例。

代码清单 34-1：string4.asm

```
; sse_string4.asm
; 查找一个字符
extern print16b
extern printf
section .data
    string1     db "qdacdekkfijlmdoza"
                db "becdfgdklkmdddaf"
                db "fffffffdedeee",10,0
    string2     db "e",0
    string3     db "a",0
    fmt         db "Find all the characters '%s' "
                db "and '%s' in:",10,0
    fmt_oc      db "I found %ld characters '%s'"
                db "and '%s'",10,0
    NL          db 10,0
```

```
section .bss
section .text
        global main
main:
push rbp
mov rbp,rsp

;打印搜索的字符
        mov     rdi, fmt
        mov     rsi, string2
        mov     rdx, string3
        xor     rax,rax
        call    printf
;打印目标字符串
        mov     rdi, string1
        xor     rax,rax
        call    printf
;搜索字符串并打印掩码
        mov     rdi, string1
        mov     rsi, string2
        mov     rdx, string3
        call    pcharsrch
;打印 string2 的出现次数
        mov     rdi, fmt_oc
        mov     rsi, rax
        mov     rdx, string2
        mov     rcx, string3
        call    printf
; 退出
leave
ret
;--------------------------------------------------------------
;用来搜索和打印掩码的函数
pcharsrch:              ;打包字符搜索
push    rbp
mov     rbp,rsp
        sub     rsp,16          ;提供用于压入 xmm1 的堆栈空间
        xor     r12,r12         ;用于出现次数的总和
        xor     rcx,rcx         ;用于发出结束信号
```

```
        xor     rbx,rbx         ;用于地址计算
        mov     rax,-16         ;字节计数,避免标志设置
;构建 xmm1,加载搜索字符
        pxor    xmm1,xmm1       ; 清除 xmm1
        pinsrb  xmm1,byte[rsi],0    ;位于索引 0 的第一个字符
        pinsrb  xmm1,byte[rdx],1    ;位于索引 1 的第二个字符
.loop:
        add     rax,16          ;避免 ZF 标识设置
        mov     rsi,16          ;如果没有终止 0,打印 16 个字节
        movdqu  xmm2,[rdi+rbx]  ;在 xmm2 中加载 16 个字节的字符串
        pcmpistrm xmm1,xmm2,40h ;'equal each'和'byte mask in xmm0'
        setz    cl              ;如果终止 0 被删除
;如果发现终止 0,确定位置
        cmp     cl,0
        je      .gotoprint      ;没有发现终止 0
;发现终止 null
;剩余少于 16 个字节
;rdi 包含字符串的地址
;rbx 包含迄今为止所处理块的字节数量
        add     rdi,rbx         ;字符串剩余部分的地址
        push    rcx             ;调用者保留(cl 在使用中)
        call    pstrlen         ;rax 返回长度
        pop     rcx             ;调用者保留
        dec     rax             ;长度,不包含 0
        mov     rsi,rax         ;剩余掩码字节的长度
;打印掩码
.gotoprint:
        call print_mask
;持续跟踪匹配的总数
        popcnt  r13d,r13d       ;计算 bit 是 1 的数量
        add     r12d,r13d       ;在 r12d 中保存出现的次数
        or      cl,cl           ; 检测到终止 0 了吗?
        jnz     .exit
        add     rbx,16          ;为接下来的 16 个字节做准备
        jmp     .loop
.exit:
        mov     rdi, NL         ;添加一个新行
        call    printf
        mov     rax,r12         ;发现的次数
```

```
        leave
        ret
;---------------------------------------------------------------
;用来查找终止 0 的函数
pstrlen:
push    rbp
mov     rbp,rsp
        sub     rsp,16              ;用来保存 xmm0
        movdqu  [rbp-16],xmm0       ;压入 xmm0
        mov     rax, -16            ;避免后面设置标志
        pxor    xmm0, xmm0          ;搜索 0(字符串末尾的终止 0)
.loop:  add rax, 16                 ;避免设置 ZF
        pcmpistri xmm0, [rdi + rax], 0x08 ;'equal each'
        jnz     .loop               ;发现 0?
        add     rax, rcx            ;rax 包含已经处理的字节
                                    ;rcx 包含在终止循环中处理的字节
        movdqu  xmm0,[rbp-16]       ;弹出 xmm0
leave
ret
;---------------------------------------------------------------
;用于打印掩码的函数
;xmm0 包含掩码
;rsi 包含要打印的 bit 的数量(16 或更少)
print_mask:
push    rbp
mov     rbp,rsp
        sub     rsp,16              ;为了保存 xmm0
        call    reverse_xmm0        ;小端格式
        pmovmskb r13d,xmm0          ;将字节掩码移到 r13d
        movdqu  [rbp-16],xmm1       ;由于 printf 压入 xmm1
        push    rdi                 ;rdi 包含 string1
        mov     edi,r13d            ;包含需要打印的掩码
        push    rdx                 ;包含掩码
        push    rcx                 ;包含字符串结束标志
        call    print16b
        pop     rcx
        pop     rdx
        pop     rdi
        movdqu  xmm1,[rbp-16]       ;弹出 xmm1
```

```
leave
ret
;------------------------------------------------------------
;用于反转和重排 xmm0 的函数
reverse_xmm0:
section .data
;用于反转的掩码
        .bytereverse  db 15,14,13,12,11,10,9,8,7,6,5,4,3,2,1,0
section .text
push rbp
mov rbp,rsp
       sub       rsp,16
       movdqu    [rbp-16],xmm2
       movdqu    xmm2,[.bytereverse]       ;在 xmm2 中加载掩码
       pshufb    xmm0,xmm2                 ;进行重排
       movdqu    xmm2,[rbp-16]             ;弹出 xmm2
leave              ;返回重排后的 xmm0
ret
```

代码清单 34-2：print16b.c

```c
// print16b.c
#include <stdio.h>
#include <string.h>
void print16b(long long n, int length){
    long long s,c;
    int i=0;
    for (c = 15; c >= 16-length; c--)
    {
        s = n >> c;
        if (s & 1)
            printf("1");
        else
            printf("0");
    }
}
```

代码清单 34-3：makefile

```
sse_string4: sse_string4.o print16b.o
```

```
        gcc -o sse_string4 sse_string4.o print16b.o -no-pie
sse_string4.o: sse_string4.asm
        nasm -f elf64 -g -F dwarf sse_string4.asm -l sse_string4.lst
printb: print16b.c
        gcc -c print16b.c
```

该程序的主要部分非常简单，但是与前面的示例一样，该程序由于要在屏幕上打印一些结果而变得复杂。我们本可以避免打印部分，而使用调试器来研究寄存器和内存中的结果。但是应付打印挑战很有趣，对吗？

输出如图 34-1 所示。

```
jo@ubuntu18:~/Desktop/Book/39 sse_string4$ ./sse_string4
Find all the characters 'e' and 'a' in:
qdacdekkfijlmdozabecdfgdklkmdddafffffffffdedeee
0010010000000000101000000000001000000000010111
I found 9 characters 'e'and 'a'
jo@ubuntu18:~/Desktop/Book/39 sse_string4$
```

图 34-1　sse_string4.asm 的输出

在示例程序中，我们将搜索字符串中的两个字符。我们提供一个字符串，恰如其分地称为 string1，并查找存储在 string2 中的字符'e'和存储在 string3 中的字符'a'。

我们使用许多函数。下面首先讨论函数 reverse_xmm0。此函数将 xmm0 作为参数，并使用重排反转字节顺序。这样，我们将能够首先从最低有效字节开始打印 xmm0，从而以小端格式打印。这就是我们在上一章中介绍重排的原因。

还有一个计算字符串长度的函数：pstrln。我们需要这个函数，因为我们将读取 16 字节块。最后一个块可能不包含 16 字节，因此对于最后一个块，我们需要确定终止 0 的位置。这将帮助我们打印一个与字符串长度相同的掩码。

自定义函数 pcharsrch 以三个字符串为参数，操作将在这个函数中执行。在该函数中，我们首先执行一些事务处理，例如初始化寄存器。寄存器 xmm1 将用作掩码；我们使用 pinsrb 指令(打包的插入字节)将要搜索的字符存储在 xmm1 中。然后开始循环，每次在 xmm2 中复制 string1 的 16 个字节，以查找字符或终止 null。使用掩码指令 pcmpistrm(将隐式长度字符串与掩码打包在一起)。pcmpistrm 指令将指定执行操作的即时控制字节作为第三个操作数，在这种情况下为"equal any"和"byte mask in xmm0"。因此，我们将寻找"equals"搜索字符串的"any"字符。对于 xmm2 中的每个匹配字符，xmm0 中与 xmm2 中的匹配字符位置对应的位将被设置为 1。pcmpistrm 指令没有将 xmm0 作为操作数，但是它被隐式使用。返回掩码将始终保存在 xmm0 中。

pcmpistri 返回的索引为 1，与 ecx 中的位置匹配。而 pcmpistrm 与其不同，将在 xmm0 中返回 16 字节块的所有匹配位置。这样一来，就可大幅减少执行步骤的数量，以查找所有匹配项。

可对 xmm0 使用位掩码或字节掩码(设置或清除控制字节中的位 6)。我们使用了字节掩码，以便可以使用调试器更轻松地读取 xmm0 寄存器，xmm0 中的两个 ff 表示一个字节，所有位均设置为 1。

研究第一个 16 字节的块之后，我们验证是否找到了终止 0，并将验证结果存储在 cl 中以备后用。我们要使用函数 print_mask 打印存储在 xmm0 中的掩码。注意在调试器中，由于使用了小端格式，字节掩码在 xmm0 中被反转了。因此，在打印之前，我们必须将其反转。这就是我们在函数 reverse_xmm0 中所做的事情。然后，我们调用 C 函数 print16b 来打印反转的掩码。但是，我们不能使用 xmm0 作为 print16b 的参数，因为在幕后 print16b 使用的是 printf，而 printf 会将 xmm0 解释为浮点值，而不是字节掩码。因此，在调用 print16b 之前，我们使用指令 pmovmksb 将 xmm0 中的位掩码传输到 r13d(意思是"移动字节掩码")。稍后将使用 r13d 进行计数；为进行打印，将其复制到 edi。我们将 xmm1 存储在堆栈中，以备后用。

我们调用 C 函数 print16b 来打印掩码。此函数以 edi(掩码)和 rsi(长度，从调用者传递)作为参数。

返回 pcharsrch 后，我们用 popcnt 指令对 r13d 中的 1 进行计数，并更新 r12d 中的计数器。我们还确定是否必须退出循环，因为在字节块中检测到终止 null。

在调用 print_mask 之前，找到终止 0 时，最后一个块的相关长度由函数 pstrlen 确定。该块的起始地址是通过将 rbx(包含先前块中已筛选的字节)添加到 rdi(字符串 1 的地址)来确定的。以 rax 返回的字符串长度用于计算 xmm0 中传递给 rsi 进行打印的剩余掩码字节数。

不要被打印的东西所湮没，重要的是理解掩码的工作原理，这是本章的主要目的。

pcmpistrm 返回的掩码可以做什么呢？由此产生的掩码可用于计算搜索参数的所有出现次数，或查找所有出现次数并将其替换为其他内容，从而创建自己的查找和替换功能。

现在，下面看看另一个搜索。

34.2 搜索某个范围内的字符

范围(range)可以是任意数量要搜索的字符，例如，所有大写字符、a 和 k 之间的所有字符、表示数字的所有字符等等。

代码清单 34-4 展示了如何在字符串中搜索大写字符。

代码清单 34-4：string5.asm

```
; sse_string5.asm
; 查找一系列字符
extern print16b
extern printf
section .data
        string1     db "eeAecdkkFijlmeoZa"
                    db "bcefgeKlkmeDad"
                    db "fdsafadfaseeE",10,0
        startrange  db "A",10,0        ;查找大写字母
        stoprange   db "Z",10,0
        NL          db 10,0
        fmt         db "Find the uppercase letters in:",10,0
        fmt_oc      db "I found %ld uppercase letters",10,0
section .bss
section .text
        global main
main:
push rbp
mov rbp,rsp
;首先打印字符串
        mov     rdi, fmt    ;title
        xor     rax,rax
        call    printf
        mov     rdi, string1        ;字符串
        xor     rax,rax
        call    printf
; 搜索字符串
        mov     rdi, string1
        mov     rsi, startrange
        mov     rdx, stoprange
        call    prangesrch
; 打印出现的次数
        mov     rdi, fmt_oc
        mov     rsi, rax
        xor     rax, rax
        call    printf
leave
```

```
ret
;-----------------------------------------------------------
;用于搜索和打印掩码的函数
prangesrch:              ;打包范围搜索
push    rbp
mov     rbp,rsp
        sub     rsp,16          ;用于压入 xmm1 的空间
        xor     r12,r12         ;用于出现的次数
        xor     rcx,rcx         ;用于发出结束信号
        xor     rbx,rbx         ;用于地址计算
        mov rax,-16             ;避免设置 ZF 标志
;构建 xmm1
        pxor    xmm1,xmm1       ; 确保一切都被清除
        pinsrb  xmm1,byte[rsi],0    ;范围从索引 0 开始
        pinsrb  xmm1,byte[rdx],1    ;范围在索引 1 结束
.loop:
        add     rax,16
        mov     rsi,16          ;如果没有终止 0, 打印 16 个字节
        movdqu  xmm2,[rdi+rbx]
        pcmpistrm xmm1,xmm2,01000100b  ; equal each|byte mask in xmm0
        setz    cl              ;删除终止 0
;如果找到终止 0, 确定位置
        cmp     cl,0
        je      .gotoprint      ;no terminating 0 found
        ;terminating null found
        ;less than 16 bytes left
        ;rdi contains address of string
        ;rbx contains #bytes in blocks handled so far
        add     rdi,rbx         ;只取字符串的尾部
        push    rcx             ;调用者保留(cl 在使用中)
        call    pstrlen         ;确定 0 的位置
        pop     rcx             ;调用者保留
        dec     rax             ;不包括终止 0 的长度
        mov     rsi,rax         ;尾部的字节
;打印掩码
.gotoprint:
        call print_mask
;持续跟踪匹配总数
```

```
        popcnt  r13d, r13d        ;bit 是 1 的数量
        add     r12d, r13d        ;在 r12 中保存出现的次数
        or      cl,cl             ;找到终止 0 了吗?
        jnz     .exit
        add     rbx,16            ;为下一区块做准备
        jmp     .loop
.exit:
        mov     rdi, NL
        call    printf
        mov     rax, r12          ;返回出现的次数
leave
ret
;-----------------------------------------------------------
pstrlen:
push    rbp
mov     rbp,rsp
        sub     rsp,16            ;用于压入 xmm0
        movdqu  [rbp-16],xmm0     ;压入 xmm0
        mov     rax, -16          ;避免后面设置 ZF 标志
        pxor    xmm0, xmm0        ;搜索 0(字符串结尾)
.loop:
        add     rax, 16           ; 当 pcmpistri 之后的 rax = 0 时避免设置 ZF
        pcmpistri xmm0, [rdi + rax], 0x08 ;'equal each'
        jnz     .loop             ;找到 0 了吗?
        add     rax, rcx          ;rax = 已经处理的字节数
                                  ;rcx = 在终止循环中处理的字节数
movdqu  xmm0,[rbp-16]             ;弹出 xmm0
leave
ret
;-----------------------------------------------------------
;用于打印掩码的函数
;xmm0 包含掩码
;rsi 包含要打印的 bit 数(16 或更少)
print_mask:
push    rbp
mov     rbp,rsp
        sub     rsp,16            ;用于保存 xmm0
        call    reverse_xmm0      ;小端格式
        pmovmskb r13d,xmm0        ;把字节掩码移到 r13d
```

```
            movdqu      [rbp-16],xmm1       ;由于 printf 压入 xmm1
            push        rdi                 ;rdi 包含 string1
            mov         edi, r13d           ;包含需要打印的掩码
            push        rdx                 ;包含掩码
            push        rcx                 ;包含字符串结束标志
            call        print16b
            pop         rcx
            pop         rdx
            pop         rdi
            movdqu      xmm1,[rbp-16]       ;弹出 xmm1
leave
ret
;---------------------------------------------------------------
;用于反转和重排 xmm0 的函数
reverse_xmm0:
section .data
;用于反转的掩码
        .bytereverse   db 15,14,13,12,11,10,9,8,7,6,5,4,3,2,1,0
section   .text
push rbp
mov  rbp,rsp
     sub    rsp,16
     movdqu [rbp-16],xmm2
     movdqu xmm2,[.bytereverse]      ;把掩码加载到 xmm2
     pshufb xmm0,xmm2                ;进行重排
     movdqu xmm2,[rbp-16]            ;弹出 xmm2
leave           ;返回重排后的 xmm0
ret
```

该程序与上一个程序几乎完全相同。我们只是给了 string2 和 string3 更有意义的名称。最重要的是，我们将传递给 pcmpistrm 的控制字节更改为 01000100b，这意味着"equal range"和"mask byte in xmm0"。

打印处理与上一节相同。

输出如图 34-2 所示。

图 34-2　sse_string5.asm 的输出

下面再看一个例子。

34.3 搜索子字符串

代码清单 34-5 展示了代码。

代码清单 34-5：string6.asm

```
; sse_string6.asm
; 查找一个子串
extern print16b
extern printf
section .data
      string1   db "a quick pink dinosour jumps over the "
                db "lazy river and the lazy dinosour "
                db "doesn't mind",10,0
      string2   db "dinosour",0
      NL        db 10,0
      fmt       db "Find the substring '%s' in:",10,0
      fmt_oc    db "I found %ld %ss",10,0

section .bss
section .text
      global main
main:
push rbp
mov rbp,rsp

;首先打印字符串
      mov    rdi, fmt
      mov    rsi, string2
      xor    rax,rax
      call   printf
      mov    rdi, string1
      xor    rax,rax
      call   printf
; 搜索字符串
      mov    rdi, string1
      mov    rsi, string2
      call   psubstringsrch
```

第 34 章 ■ SSE 字符串掩码

```
;打印子串出现的次数
        mov     rdi, fmt_oc
        mov     rsi, rax
        mov     rdx, string2
        call    printf
leave
ret
;------------------------------------------------------------
;用于搜索子串并打印掩码的函数

psubstringsrch:             ;打包子串搜索
push    rbp
mov     rbp,rsp
        sub     rsp,16          ;为保存 xmm1 预留的空间
        xor     r12,r12         ;管理发现的总数
        xor     rcx,rcx         ;用于发出结束信号
        xor     rbx,rbx         ;用于地址计算
        mov rax,-16             ;避免设置 ZF 标志
;构建 xmm1, 加载子串
        pxor xmm1,xmm1
        movdqu  xmm1,[rsi]
.loop:
        add rax,16          ;避免设置 ZF 标志
        mov rsi,16          ;如果没有 0, 打印 16 个字节
        movdqu  xmm2,[rdi+rbx]
        pcmpistrm xmm1,xmm2,01001100b ;'equal ordered'|'byte mask in xmm0'
        setz cl         ;删除终止 0

;如果发现终止 0, 确定位置
        cmp cl,0
        je      .gotoprint      ;没有发现终止 0
        ;发现终止 null
        ;剩余少于 16 字节
        ;rdi 包含字符串的地址
        ;rbx 包含迄今为止处理的块中的字节数
        add     rdi,rbx         ;只取字符串的末尾
        push    rcx             ;调用者保留(cl 在使用中)
        call    pstrlen         ;rax 返回 0 的位置
        push    rcx             ;调用者保留(cl 在使用中)
        dec     rax             ;不包括终止 0 的长度
```

```
            mov     rsi,rax         ;剩余字节的长度
;打印掩码
.gotoprint:
            call    print_mask
;管理匹配的总数
            popcnt  r13d,r13d       ;bit 是 1 的数量
            add     r12d,r13d       ;在 r12 中保存出现的次数
            or      cl,cl           ;检测到终止 0 了吗?
            jnz     .exit
            add     rbx,16          ;为下一区块做准备
            jmp     .loop
.exit:
            mov     rdi, NL
            call    printf
            mov     rax, r12        ;返回出现的次数
leave
ret
;----------------------------------------------------------------
pstrlen:
push    rbp
mov     rbp,rsp
            sub     rsp,16          ;用于压入 xmm0
            movdqu  [rbp-16],xmm0   ;压入 xmm0
            mov     rax, -16        ;避免以后设置 ZF 标志
            pxor    xmm0, xmm0      ;搜索 0(字符串结尾)
.loop:
            add     rax, 16         ;当 pcmpistri 之后是 rax = 0 时避免设置 ZF
            pcmpistri xmm0, [rdi + rax], 0x08 ;'equal each'
            jnz     .loop           ;发现 0 了吗?
            add     rax, rcx        ;rax = 已经处理的字节数
                                    ;rcx = 在终止循环中已经处理的字节数
            movdqu xmm0,[rbp-16]    ;弹出 xmm0
leave
ret
;----------------------------------------------------------------
;用于打印掩码的函数
;xmm0 包含掩码
;rsi 包含要打印的位数(16 或更少)
```

```
print_mask:
push    rbp
mov     rbp,rsp
        sub         rsp,16                  ;为保存 xmm0
        call        reverse_xmm0            ;小端格式
        pmovmskb    r13d,xmm0               ;将字节掩码移到 edx
        movdqu      [rbp-16],xmm1           ;由于 printf 压入 xmm1
        push        rdi                     ;rdi 包含 string1
        mov         edi,r13d                ;包含需要打印的掩码
        push        rdx                     ;包含掩码
        push        rcx                     ;包含字符串结束标志
        call        print16b
        pop         rcx
        pop         rdx
        pop         rdi
        movdqu      xmm1,[rbp-16]           ;弹出 xmm1
leave
ret
;-------------------------------------------------------------
;用于反转和重排 xmm0 的函数
reverse_xmm0:
section .data
;用于反转的掩码
.bytereverse db 15,14,13,12,11,10,9,8,7,6,5,4,3,2,1,0
section .text
push    rbp
mov     rbp,rsp
        sub         rsp,16
        movdqu      [rbp-16],xmm2
        movdqu      xmm2,[.bytereverse]     ;在 xmm2 中加载掩码
        pshufb      xmm0,xmm2               ;进行重排
        movdqu      xmm2,[rbp-16]           ;弹出 xmm2
leave                                       ;返回重排后的 xmm0
ret
```

我们使用了与以前几乎相同的代码，只是更改了字符串，并且控制字节包含 "equal ordered"和"byte mask in xmm0"。

输出如图 34-3 所示。

```
jo@ubuntu18:~/Desktop/Book/41 sse_string6$ ./sse_string6
Find the substring 'dinosour' in:
a quick pink dinosour jumps over the lazy river and the lazy dinosour doesn't mind
0000000000000100000000000000000000000000000000000000000100000000000000000000
I found 2 dinosours
jo@ubuntu18:~/Desktop/Book/41 sse_string6$
```

图 34-3　sse_string6.asm 的输出

34.4　小结

本章内容：
- 使用字符串掩码
- 搜索字符、范围和子字符串
- 从 xmm 寄存器打印掩码

第 35 章

AVX

高级矢量扩展(AVX)是 SSE 的扩展。SSE 提供 16 个具有 128 位的 xmm 寄存器，而 AVX 提供 16 个具有 256 位的 ymm 寄存器。每个 ymm 寄存器的下半部分实际上是对应的 xmm 寄存器。xmm 寄存器是 ymm 寄存器的别名。AVX-512 是进一步的扩展，提供 32 个具有 128 位的 zmm 寄存器。

除了这些寄存器，AVX 还扩展了 SSE 指令并提供了一系列额外的新指令。完成本书的 SSE 章节后，你会发现查阅大量 SSE 和 AVX 指令也不困难。

在本章中，我们将首先说明处理器支持哪个 AVX 版本，然后将展示一个示例程序。

35.1 测试是否支持 AVX

代码清单 35-1 展示了一个程序，可用来确定你的 CPU 是否支持 AVX。

代码清单 35-1：cpu_avx.asm

```
; cpu_avx.asm
extern printf
section .data
        fmt_noavx     db "This cpu does not support AVX.",10,0
        fmt_avx       db "This cpu supports AVX.",10,0
        fmt_noavx2    db "This cpu does not support AVX2.",10,0
        fmt_avx2      db "This cpu supports AVX2.",10,0
        fmt_noavx512  db "This cpu does not support AVX-512.",10,0
        fmt_avx512    db "This cpu supports AVX-512.",10,0
section .bss
```

```
section .text
    global main
main:
push rbp
mov rbp,rsp
    call cpu_sse      ; 如果支持AVX,则在rax中返回1;否则返回0
leave
ret

cpu_sse:
push rbp
mov rbp,rsp
; 测试avx
    mov   eax,1       ; 请求CPU功能标志
    cpuid
    mov   eax,28      ; 在ecx中测试位28
    bt    ecx,eax
    jnc   no_avx
    xor   rax,rax
    mov   rdi,fmt_avx
    call  printf
;测试avx2
    mov   eax,7       ; 请求CPU功能标志
    mov   ecx,0
    cpuid
    mov   eax,5       ; 在ebx中测试位5
    bt    ebx,eax
    jnc   the_exit
    xor   rax,rax
    mov   rdi,fmt_avx2
    call  printf
;avx512基础测试
    mov   eax,7       ; 请求CPU功能标志
    mov   ecx,0
    cupid
    mov   eax,16      ; 在ebx中测试位16
    bt    ebx,eax
    jnc   no_avx512
    xor   rax,rax
```

```
        mov    rdi,fmt_avx512
        call   printf
        jmp    the_exit
no_avx:
        mov    rdi,fmt_noavx
        xor    rax,rax
        call   printf           ; 如果 AVX 不可用，则显示消息
        xor    rax,rax          ; 如果不支持 AVX，则返回 0
        jmp    the_exit         ; 退出
no_avx2:
        mov    rdi,fmt_noavx2
        xor    rax,rax
        call   printf           ; 如果 AVX 不可用，则显示消息
        xor    rax,rax          ; 如果不支持 AVX，则返回 0
        jmp    the_exit         ; 退出
no_avx512:
        mov    rdi,fmt_noavx512
        xor    rax,rax
        call   printf           ; 如果 AVX 不可用，则显示消息
        xor    rax,rax          ; 如果不支持 AVX，则返回 0
        jmp    the_exit         ; 退出
the_exit:
leave
ret
```

该程序类似于我们用于测试是否支持 SSE 的程序，但是我们现在必须查找 AVX 标志。这里没有什么特别之处，你可以在英特尔手册第 2 卷的 cpuid 一节中找到有关要使用哪些寄存器以及可以检索哪些信息的详细内容。

输出如图 35-1 所示。

```
jo@ubuntu18:~/Desktop/Book/42 cpu_avx$ ./cpu_avx
This cpu supports AVX.
This cpu supports AVX2.
This cpu does not support AVX-512.
jo@ubuntu18:~/Desktop/Book/42 cpu_avx$
```

图 35-1　cpu_avx.asm 的输出

35.2　AVX 程序示例

代码清单 35-2 改编自第 28 章中的 SSE 未对齐示例。

代码清单 35-2：avx_unaligned.asm

```
; avx_unaligned.asm
extern printf
section .data
        spvector1   dd 1.1
                    dd 2.1
                    dd 3.1
                    dd 4.1
                    dd 5.1
                    dd 6.1
                    dd 7.1
                    dd 8.1

        spvector2   dd 1.2
                    dd 1.2
                    dd 3.2
                    dd 4.2
                    dd 5.2
                    dd 6.2
                    dd 7.2
                    dd 8.2

        dpvector1   dq 1.1
                    dq 2.2
                    dq 3.3
                    dq 4.4

        dpvector2   dq 5.5
                    dq 6.6
                    dq 7.7
                    dq 8.8

        fmt1 db "Single Precision Vector 1:",10,0
        fmt2 db 10,"Single Precision Vector 2:",10,0
        fmt3 db 10,"Sum of Single Precision Vector 1 and Vector 2:",10,0
        fmt4 db 10,"Double Precision Vector 1:",10,0
        fmt5 db 10,"Double Precision Vector 2:",10,0
        fmt6 db 10,"Sum of Double Precision Vector 1 and Vector 2:",10,0

section .bss
        spvector_res  resd 8
```

```
            dpvector_res   resq 4
section .text
        global main
main:
push rbp
mov rbp,rsp
;单精度浮点向量
;在寄存器 ymm0 中加载 vector1
        vmovups    ymm0, [spvector1]
;提取 ymm0
        vextractf128   xmm2,ymm0,0     ;ymm0 的第一部分
        vextractf128   xmm2,ymm0,1     ;ymm0 的第二部分
;在寄存器 ymm1 中加载 vector2
        vmovups    ymm1, [spvector2]
;提取 ymm1
        vextractf128   xmm2,ymm1,0
        vextractf128   xmm2,ymm1,1
;添加两个单精度浮点向量
        vaddps     ymm2,ymm0,ymm1
        vmovups    [spvector_res],ymm2
;打印向量
        mov   rdi,fmt1
        call printf
        mov   rsi,spvector1
        call printspfpv
        mov   rdi,fmt2
        call printf
        mov   rsi,spvector2
        call printspfpv
        mov   rdi,fmt3
        call printf
        mov   rsi,spvector_res
        call printspfpv
;双精度浮点向量
;在寄存器 ymm0 中加载 vector1
        vmovups   ymm0, [dpvector1]
;提取 ymm0
        vextractf128   xmm2,ymm0,0     ;ymm0 的第一部分
        vextractf128   xmm2,ymm0,1     ;ymm0 的第二部分
```

```asm
;在寄存器 ymm1 中加载 vector2
        vmovups  ymm1,[dpvector2]
;提取 ymm1
        vextractf128  xmm2,ymm1,0
        vextractf128  xmm2,ymm1,1
;添加两个双精度浮点向量
        vaddpd  ymm2,ymm0,ymm1
        vmovupd [dpvector_res],ymm2
;打印向量
        mov  rdi,fmt4
        call printf
        mov  rsi,dpvector1
        call printdpfpv
        mov  rdi,fmt5
        call printf
        mov  rsi,dpvector2
        call printdpfpv
        mov  rdi,fmt6
        call printf
        mov  rsi,dpvector_res
        call printdpfpv
leave
ret

printspfpv:
section .data
        .NL   db 10,0
        .fmt1 db "%.1f, ",0
section .text
push  rbp
mov   rbp,rsp
        push rcx
        push rbx
        mov rcx,8
        mov rbx,0
        mov rax,1
.loop:
        movss    xmm0,[rsi+rbx]
        cvtss2sd xmm0,xmm0
```

第35章 AVX

```
            mov     rdi,.fmt1
            push    rsi
            push    rcx
            call    printf
            pop     rcx
            pop     rsi
            add     rbx,4
            loop    .loop
            xor     rax,rax
            mov     rdi,.NL
            call    printf
            pop     rbx
            pop     rcx
    leave
    ret
    printdpfpv:
    section .data
            .NL db 10,0
            .fmt db "%.1f, %.1f, %.1f, %.1f",0
    section .text
    push    rbp
    mov     rbp,rsp
            mov rdi,.fmt
            mov rax,4      ; 四个浮点数
            call printf
            mov rdi,.NL
            call printf
    leave
    ret
```

在此程序中，我们使用 256 位 ymm 寄存器和一些新指令。例如，我们使用 vmovups 将未对齐的数据放入 ymm 寄存器。我们使用 SASM 查看寄存器。在 vmovups 指令之后，ymm0 包含以下内容：

{0x40833333404666664006666663f8ccccd,0x4101999a40e3333340c3333340a33333}

转换成十进制的样式如下：

{4.1 3.1 2.1 1.1, 8.1 7.1 6.1 5.1}

看看这些值存储在哪里，这可能会让人困惑。

出于演示目的，我们从 ymm 寄存器中提取数据，然后使用 vextractf128 一次性将 128 位的打包浮点值从 ymm0 放入 xmm2。你可以使用 extractps 进一步提取浮点值并将其存储在通用寄存器中。

新的指令有三个操作数，如下所示：

```
vaddps ymm2,ymm0,ymm1
```

将 ymm1 添加到 ymm0 并将结果存储在 ymm2 中。

print 函数只需要将内存中的值加载到 xmm 寄存器中，在需要时将单精度转换为双精度，然后调用 printf。

输出如图 35-2 所示。

```
jo@ubuntu18:~/Desktop/Book/43 avx_unaligned$ ./avx_unaligned
Single Precision Vector 1:
1.1,  2.1,  3.1,  4.1,  5.1,  6.1,  7.1,  8.1,

Single Precision Vector 2:
1.2,  1.2,  3.2,  4.2,  5.2,  6.2,  7.2,  8.2,

Sum of Single Precision Vector 1 and Vector 2:
2.3,  3.3,  6.3,  8.3,  10.3, 12.3, 14.3, 16.3,

Double Precision Vector 1:
1.1,  2.2,  3.3,  4.4

Double Precision Vector 2:
5.5,  6.6,  7.7,  8.8

Sum of Double Precision Vector 1 and Vector 2:
6.6,  8.8,  11.0, 13.2
jo@ubuntu18:~/Desktop/Book/43 avx_unaligned$
```

图 35-2　avx_unaligned.asm 的输出

35.3　小结

本章内容：

- 如何确定 CPU 支持 AVX
- AVX 使用 16 个 256 位的 ymm 寄存器
- 128 位 xmm 寄存器是 ymm 寄存器的别名
- 如何从 ymm 寄存器中提取值

第 36 章

AVX 矩阵运算

除了总结一些可能有趣的 AVX 指令,我们来看一些使用 AVX 执行的矩阵运算。这一章内容较多,包含几页代码。虽然其中的很多你都很熟悉,但这里将介绍几个新指令。

我们将展示矩阵乘法和矩阵求逆。在下一章中,将介绍如何转置矩阵。

36.1 矩阵代码示例

代码清单 36-1 展示了示例代码。

代码清单 36-1:矩阵 4x4.asm

```
; matrix4x4.asm
extern printf

section .data
        fmt0    db 10,"4x4 DOUBLE PRECISION FLOATING POINT MATRICES",10,0
        fmt1    db 10,"This is matrixA:",10,0
        fmt2    db 10,"This is matrixB:",10,0
        fmt3    db 10,"This is matrixA x matrixB:",10,0
        fmt4    db 10,"This is matrixC:",10,0
        fmt5    db 10,"This is the inverse of matrixC:",10,0
        fmt6    db 10,"Proof: matrixC x inverse =",10,0
        fmt7    db 10,"This is matrixS:",10,0
        fmt8    db 10,"This is the inverse of matrixS:",10,0
        fmt9    db 10,"Proof: matrixS x inverse =",10,0
        fmt10   db 10,"This matrix is singular!",10,10,0
```

```
        align 32
    matrixA  dq 1., 3., 5., 7.
             dq 9., 11., 13., 15.
             dq 17., 19., 21., 23.
             dq 25., 27., 29., 31.

    matrixB  dq 2., 4., 6., 8.
             dq 10., 12., 14., 16.
             dq 18., 20., 22., 24.
             dq 26., 28., 30., 32.

    matrixC  dq 2., 11., 21., 37.
             dq 3., 13., 23., 41.
             dq 5., 17., 29., 43.
             dq 7., 19., 31., 47.

    matrixS  dq 1., 2., 3., 4.
             dq 5., 6., 7., 8.
             dq 9., 10., 11., 12.
             dq 13., 14., 15., 16.

section .bss
    alignb 32
    product  resq 16
    inverse  resq 16
section .text
    global main
main:
push rbp
mov rbp,rsp
; 打印标题
    mov rdi, fmt0
    call printf
; 打印 matrixA
    mov rdi,fmt1
    call printf
    mov rsi,matrixA
    call printm4x4
; 打印 matrixB
    mov rdi,fmt2
    call printf
```

```
        mov rsi,matrixB
        call printm4x4
; 计算 matrixA 和 matrixB 的乘积
        mov rdi,matrixA
        mov rsi,matrixB
        mov rdx,product
        call multi4x4
; 打印乘积
        mov rdi,fmt3
        call printf
        mov rsi,product
        call printm4x4

; 打印 matrixC
        mov rdi,fmt4
        call printf
        mov rsi,matrixC
        call printm4x4
; 计算 matrixC 的逆
        mov rdi,matrixC
        mov rsi,inverse
        call inverse4x4
        cmp rax,1
        je singular
; 打印逆
        mov rdi,fmt5
        call printf
        mov rsi,inverse
        call printm4x4
; 证明 matrixC 乘以逆
        mov rsi,matrixC
        mov rdi,inverse
        mov rdx,product
        call multi4x4
; 打印证明结果
        mov rdi,fmt6
        call printf
        mov rsi,product
        call printm4x4
```

```
        ; 奇异矩阵
        ; 打印 matrixS
                mov rdi,fmt7
                call printf
                mov rsi,matrixS
                call printm4x4
        ; 计算 matrixS 的逆
                mov rdi,matrixS
                mov rsi,inverse
                call inverse4x4
                cmp rax,1
                je  singular
        ; 打印逆
                mov rdi,fmt8
                call printf
                mov rsi,inverse
                call printm4x4
        ; 证明 matrixS 和逆相乘
                mov rsi,matrixS
                mov rdi,inverse
                mov rdx,product
                call multi4x4
        ; 打印证明结果
                mov rdi,fmt9
                call printf
                mov rsi,product
                call printm4x4
                jmp exit
singular:
        ; 打印错误
                mov rdi,fmt10
                call printf
exit:
leave
ret

inverse4x4:
section .data
        align 32
```

```
            .identity    dq   1., 0., 0., 0.
                         dq   0., 1., 0., 0.
                         dq   0., 0., 1., 0.
                         dq   0., 0., 0., 1.

            .minus_mask  dq   8000000000000000h
            .size        dq   4          ;4 x 4 矩阵
            .one         dq   1.0
            .two         dq   2.0
            .three       dq   3.0
            .four        dq   4.0
section .bss
      alignb 32
            .matrix1   resq 16      ;中间矩阵
            .matrix2   resq 16      ;中间矩阵
            .matrix3   resq 16      ;中间矩阵
            .matrix4   resq 16      ;中间矩阵
            .matrixI   resq 16

            .mxcsr     resd 1       ;用于检查零除法

section .text
push rbp
mov rbp,rsp
            push rsi                ;保存逆矩阵的地址
            vzeroall                ;清除所有 ymm 寄存器
;  计算中间矩阵
;  计算中间矩阵 matrix2
;  rdi 包含原始矩阵的地址
            mov rsi,rdi
            mov rdx,.matrix2
            push rdi
            call multi4x4
            pop rdi

;  计算中间矩阵 matrix3
            mov rsi,.matrix2
            mov rdx,.matrix3
            push rdi
            call multi4x4
```

```
            pop rdi
    ; 计算中间矩阵 matrix4
            mov rsi,.matrix3
            mov rdx,.matrix4
            push rdi
            call multi4x4
            pop rdi
    ;计算迹(trace)
    ;计算 trace1
            mov rsi,[.size]
            call vtrace
            movsd  xmm8,xmm0         ;trace1 在 xmm8 中
    ;计算 trace2
            push rdi             ; 保存原始矩阵的地址
            mov rdi,.matrix2
            mov rsi,[.size]
            call vtrace
            movsd  xmm9,xmm0         ;trace2 在 xmm9 中
    ;计算 trace3
            mov rdi,.matrix3
            mov rsi,[.size]
            call vtrace
            movsd  xmm10,xmm0        ;trace3 在 xmm10 中
    ;计算 trace4
            mov rdi,.matrix4
            mov rsi,[.size]
            call vtrace
            movsd  xmm11,xmm0        ;trace 4 在 xmm11 中
    ; 计算系数
    ; 计算系数 p1
    ; p1 = -s1
            vxorpd  xmm12,xmm8,[.minus_mask]  ;p1 在 xmm12 中
    ; 计算系数 p2
    ; p2 = -1/2 * (p1 * s1 + s2)
            movsd    xmm13,xmm12     ;把 p1 复制到 xmm13
            vfmadd213sd  xmm13,xmm8,xmm9  ;xmm13=xmm13*xmm8+xmm9
            vxorpd   xmm13,xmm13,[.minus_mask]
```

```
        divsd     xmm13,[.two]       ;除以2，p2 在 xmm13 中
; 计算系数 p3
; p3 = -1/3 * (p2 * s1 + p1 * s2 + s3)
        movsd     xmm14,xmm12        ;把 p1 复制到 xmm14
        vfmadd213sd xmm14,xmm9,xmm10  ;p1*s2+s3;xmm14=xmm14*xmm9+xmm10
        vfmadd231sd xmm14,xmm13,xmm8  ;xmm14+p2*s1;xmm14=xmm14+xmm13*xmm8
        vxorpd xmm14,xmm14,[.minus_mask]
        divsd xmm14,[.three]         ;p3 在 xmm14 中
; 计算系数 p4
; p4 = -1/4 * (p3 * s1 + p2 * s2 + p1 * s3 + s4)
        movsd xmm15,xmm12            ;把 p1 复制到 xmm15
        vfmadd213sd xmm15,xmm10,xmm11 ;p1*s3+s4;xmm15=xmm15*xmm10+xmm11
        vfmadd231sd xmm15,xmm13,xmm9  ;xmm15+p2*s2;xmm15=xmm15+xmm13*xmm9
        vfmadd231sd xmm15,xmm14,xmm8  ;xmm15+p3*s1;xmm15=xmm15+xmm14*xmm8
        vxorpd xmm15,xmm15,[.minus_mask]
        divsd xmm15,[.four]          ;p4 在 xmm15 中
;用适当的系数乘以矩阵
        mov rcx,[.size]
        xor rax,rax

        vbroadcastsd   ymm1,xmm12    ; p1
        vbroadcastsd   ymm2,xmm13    ; p2
        vbroadcastsd   ymm3,xmm14    ; p3


        pop rdi          ; 恢复原矩阵的地址

.loop1:
        vmovapd   ymm0,[rdi+rax]
        vmulpd    ymm0,ymm0,ymm2
        vmovapd   [.matrix1+rax],ymm0

        vmovapd   ymm0,[.matrix2+rax]
        vmulpd    ymm0,ymm0,ymm1
        vmovapd   [.matrix2+rax],ymm0

        vmovapd   ymm0,[.identity+rax]
        vmulpd    ymm0,ymm0,ymm3
        vmovapd   [.matrixI+rax],ymm0

        add       rax,32
```

```
            loop    .loop1
;将四个矩阵相加并乘以-1/p4
            mov     rcx,[.size]
            xor     rax,rax
;计算-1/p4
            movsd   xmm0, [.one]
            vdivsd  xmm0,xmm0,xmm15      ;1/p4
;检查除以零
            stmxcsr [.mxcsr]
            and     dword[.mxcsr],4
            jnz     .singular
; 没有除以零
            pop           rsi            ;逆矩阵的地址
            vxorpd        xmm0,xmm0,[.minus_mask] ;-1/p4
            vbroadcastsd  ymm2,xmm0
;遍历所有行
.loop2:
            ;添加行
            vmovapd   ymm0,[.matrix1+rax]
            vaddpd    ymm0, ymm0, [.matrix2+rax]
            vaddpd    ymm0, ymm0, [.matrix3+rax]
            vaddpd    ymm0, ymm0, [.matrixI+rax]
            vmulpd    ymm0,ymm0,ymm2      ;将行乘以-1/p4
            vmovapd   [rsi+rax],ymm0
            add       rax,32
            loop      .loop2

            xor       rax,rax             ;返回0,没有错误
leave
ret

.singular:
            mov rax,1            ;返回1,奇异矩阵
leave
ret
;--------------------------------------------------------
; 迹计算
vtrace:
```

```
push rbp
mov rbp,rsp
;在内存中构建矩阵
       vmovapd   ymm0, [rdi]
       vmovapd   ymm1, [rdi+32]
       vmovapd   ymm2, [rdi+64]
       vmovapd   ymm3, [rdi+96]
       vblendpd  ymm0,ymm0,ymm1,0010b
       vblendpd  ymm0,ymm0,ymm2,0100b
       vblendpd  ymm0,ymm0,ymm3,1000b
       vhaddpd   ymm0,ymm0,ymm0
       vpermpd   ymm0,ymm0,00100111b
       haddpd    xmm0,xmm0
leave
ret
;--------------------------------------------------
printm4x4:
section .data
       .fmt db "%f",9,"%f",9, "%f",9,"%f",10,0
section .text
push rbp
mov rbp,rsp
push rbx          ;被调用者保留
push r15          ;被调用者保留
     mov rdi,.fmt
     mov rcx,4
     xor   rbx,rbx       ;行计数器
.loop:
     movsd  xmm0, [rsi+rbx]
     movsd  xmm1, [rsi+rbx+8]
     movsd  xmm2, [rsi+rbx+16]
     movsd  xmm3, [rsi+rbx+24]
     mov    rax,4        ;四个浮点数
     push   rcx          ;调用者保留
     push   rsi          ;调用者保留
     push   rdi          ;调用者保留
           ;如有必要,对齐堆栈
           xor r15,r15
           test rsp,0xf         ;最后一个字节是8 (没有对齐)?
```

```asm
                setnz   r15b            ;如果没有对齐则进行设置
                shl     r15,3           ;乘以8
                sub     rsp,r15         ;减去0或8
        call    printf
                add     rsp,r15         ;加0或8恢复rsp
                pop     rdi
                pop     rsi
                pop     rcx
                add     rbx,32          ;下一行
                loop    .loop
pop r15
pop rbx
leave
ret
;------------------------------------------------------
multi4x4:
push    rbp
mov     rbp,rsp

        xor     rax,rax
        mov     rcx,4
        vzeroall                ;ymm全部置0
.loop:
        vmovapd         ymm0, [rsi]

        vbroadcastsd    ymm1,[rdi+rax]
        vfmadd231pd     ymm12,ymm1,ymm0

        vbroadcastsd    ymm1,[rdi+32+rax]
        vfmadd231pd     ymm13,ymm1,ymm0

        vbroadcastsd    ymm1,[rdi+64+rax]
        vfmadd231pd     ymm14,ymm1,ymm0

        vbroadcastsd    ymm1,[rdi+96+rax]
        vfmadd231pd     ymm15,ymm1,ymm0

        add     rax,8           ;一个元素有8个字节，64位
        add     rsi,32          ;每一行有32个字节，256位

        loop    .loop
```

```
    ;把结果逐行移到内存
        vmovapd     [rdx],      ymm12
        vmovapd     [rdx+32],   ymm13
        vmovapd     [rdx+64],   ymm14
        vmovapd     [rdx+96],   ymm15
        xor         rax,rax                 ;返回值
leave
ret
```

这段代码的有趣部分位于函数中。主要功能是初始化程序、调用函数和打印。本例中使用的矩阵是 4×4 双精度浮点矩阵。注意矩阵的 32 字节对齐；在 AVX 中，我们使用 ymm 寄存器，大小为 32 字节。我们将逐一分析程序的功能。

36.2　矩阵打印：printm4x4

我们一次将矩阵读入四个 xmm 寄存器，然后将多个寄存器压入堆栈。这些寄存器将被 printf 修改，因此必须保留。然后在 16 字节的边界上对齐堆栈。因为一般情况下，rsp 是在 8 字节边界上对齐的。为在 16 字节的边界上对齐堆栈，不能使用第 16 章中 and 指令的"把戏"。这是因为对于 and 指令，我们不知道是否会更改 rsp。而且需要正确的堆栈指针，因为在 printf 之后弹出被压入的寄存器。如果更改了 rsp，则需要在弹出之前将其恢复为之前的值；否则，错误的值将从堆栈中弹出。如果未更改 rsp，则不必进行调整。

我们将使用 test 指令和 0xf 来验证堆栈是否对齐。如果 rsp 的最后一个十六进制数字是 0，那么 rsp 是 16 字节对齐的。如果最后一个数字包含 0 以外的任何内容，那么最后半个字节将至少有一个位设置为 1。test 指令类似于 and 指令。如果 rsp 的最后半个字节将一个或多个位设置为 1，则比较结果将为非零，并将清除零标志 ZF。setnz(set-if-non-zero)指令读取零标志(ZF)，如果未设置 ZF，则 setnz 将 0000 0001 放入 r15b。如果发生这种情况，则意味着 rsp 不是 16 字节对齐的，将其减去 8 并放在 16 字节边界上。将 r15b 左移三次以获得十进制值 8 并进行减法运算。执行 printf 后，通过将 r15 加回到 rsp 来恢复正确的堆栈地址。也就是说，如果必须对齐，则加 8；如果不需要对齐，则加 0。堆栈就是对齐之前的位置，然后弹出寄存器。

36.3　矩阵乘法：multi4x4

在示例代码和说明中使用以下两个矩阵。

$$A = \begin{bmatrix} 1 & 3 & 5 & 7 \\ 9 & 11 & 13 & 15 \\ 17 & 19 & 21 & 23 \\ 25 & 27 & 29 & 31 \end{bmatrix} \quad B = \begin{bmatrix} 2 & 4 & 6 & 8 \\ 10 & 12 & 14 & 16 \\ 18 & 20 & 22 & 24 \\ 26 & 28 & 30 & 32 \end{bmatrix}$$

如果学习过一点线性代数，则可能学习过矩阵乘法。要获得矩阵 $C=AB$ 的元素 $c11$，计算公式如下所示：

$$a_{11}b_{11} + a_{12}b_{21} + a_{13}b_{31} + a_{14}b_{41}$$

在我们的示例中，它看起来像这样：

$$1\times 2 + 3\times 10 + 5\times 18 + 7\times 26 = 304$$

再举一个示例，元素 $c32$ 的计算如下：

$$a_{31}b_{12} + a_{32}b_{22} + a_{33}b_{32} + a_{34}b_{42}$$

在我们的示例中，它看起来像这样：

$$17\times 4 + 19\times 12 + 21\times 20 + 23\times 28 = 1360$$

这对于手动计算非常有效；但我们将使用一种更适合计算机的方法。我们将使用 ymm 寄存器来保持运行总和并在后续循环中更新总和。这里利用了 AVX 指令的强大功能。

首先用 vzeroall 清除所有 ymm 寄存器。然后循环四次，对 matrixB 中的每行都循环一次。将来自 matrixB 的四个双精度值的行加载到 ymm0 中。然后，将来自顺序选择的 matrixA 列的值广播到 ymm1 中。寄存器 rax 用作列计数器，列值位于偏移量 0、32、64 和 96 处。广播表示所有四个 quadwords(每个 8 字节)都将包含该值。然后，将 ymm1 中的值与 ymm0 中的值相乘并添加到 ymm12 中。乘法和加法使用一条称为 vfmadd231pd(意思是"矢量融合乘法加法压缩双精度")的指令完成。231 指示如何使用寄存器。vfmadd (132, 213, 231)有多种变体，并且有双精度和单精度的变体。我们使用 231，这意味着将第二个操作数与第三个操作数相乘，添加到第一个操作数上，然后将结果放入第一个操作数中。对 matrixA 列的每个列值都执行此操作，然后继续迭代。matrixB 的下一行将被加载，计算将重新开始。

使用你最喜欢的调试器运行该程序。看一下寄存器 ymm12、ymm13、ymm14 和 ymm15 如何保存运行总和，最后给出乘积。调试器可能会以十六进制和小端格式给出 ymm 寄存器中的值。为简单起见，下面列出每个步骤的细节。

	rdi					rsi			
	32 字节					32 字节			
	8 字节	8 字节	8 字节	8 字节		8 字节	8 字节	8 字节	8 字节
0~31	1	3	5	7	0~31	2	4	6	8
32~63	9	11	13	15	32~63	10	12	14	16
64~95	17	19	21	23	64~95	18	20	22	24
96~127	25	27	29	31	96~127	26	28	30	32

这是第一个循环：

vmovapd ymm0, [rsi]	ymm0	2	4	6	8
vbroadcastsd ymm1,[rdi+0]	ymm1	1	1	1	1
vfmadd231pd ymm12,ymm1,ymm0	ymm12	2	4	6	8
vbroadcastsd ymm1,[rdi+32+0]	ymm1	9	9	9	9
vfmadd231pd ymm13,ymm1,ymm0	ymm13	18	36	54	72
vbroadcastsd ymm1,[rdi+64+0]	ymm1	17	17	17	17
vfmadd231pd ymm14,ymm1,ymm0	ymm14	34	68	102	136
vbroadcastsd ymm1,[rdi+96+0]	ymm1	25	25	25	25
vfmadd231pd ymm15,ymm1,ymm0	ymm15	50	100	150	200

这是第二个循环：

vmovapd ymm0, [rsi+32]	ymm0	10	12	14	16
vbroadcastsd ymm1,[rdi+8]	ymm1	3	3	3	3
vfmadd231pd ymm12,ymm1,ymm0	ymm12	3	40	48	56
vbroadcastsd ymm1,[rdi+32+8]	ymm1	11	11	11	11
vfmadd231pd ymm13,ymm1,ymm0	ymm13	128	168	208	248
vbroadcastsd ymm1,[rdi+64+8]	ymm1	19	19	19	19
vfmadd231pd ymm14,ymm1,ymm0	ymm14	224	296	368	440
vbroadcastsd ymm1,[rdi+96+8]	ymm1	27	27	27	27
vfmadd231pd ymm15,ymm1,ymm0	ymm15	320	424	528	632

这是第三个循环：

vmovapd ymm0, [rsi+32+32]	ymm0	18	20	22	24
vbroadcastsd ymm1,[rdi+8+8]	ymm1	5	5	5	5
vfmadd231pd ymm12,ymm1,ymm0	ymm12	122	140	158	176

vbroadcastsd ymm1,[rdi+32+8+8]	ymm1	13	13	13	13
vfmadd231pd ymm13,ymm1,ymm0	ymm13	362	428	494	560
vbroadcastsd ymm1,[rdi+64+8+8]	ymm1	21	21	21	21
vfmadd231pd ymm14,ymm1,ymm0	ymm14	602	716	830	944
vbroadcastsd ymm1,[rdi+96+8+8]	ymm1	29	29	29	29
vfmadd231pd ymm15,ymm1,ymm0	ymm15	842	1004	1166	1328

这是第四个循环：

vmovapd ymm0, [rsi+32+32+32]	ymm0	26	28	30	32
vbroadcastsd ymm1,[rdi+8+8+8]	ymm1	7	7	7	7
vfmadd231pd ymm12,ymm1,ymm0	ymm12	304	336	368	400
vbroadcastsd ymm1,[rdi+32+8+8+8]	ymm1	15	15	15	15
vfmadd231pd ymm13,ymm1,ymm0	ymm13	752	848	944	1040
vbroadcastsd ymm1,[rdi+64+8+8+8]	ymm1	23	23	23	23
vfmadd231pd ymm14,ymm1,ymm0	ymm14	1200	1360	1520	1680
vbroadcastsd ymm1,[rdi+96+8+8+8]	ymm1	31	31	31	31
vfmadd231pd ymm15,ymm1,ymm0	ymm15	1648	1872	2096	2320

36.4 矩阵求逆：Inverse4x4

数学家们开发了一系列算法来有效地计算矩阵的逆。这不是我们的目的，我们只想展示如何使用 AVX。

我们将使用一种基于特征多项式的 Cayley-Hamilton 定理的方法。这里有一个有趣的站点，它包含关于特征多项式的更多信息：http://www.mcs.csueastbay.edu/~malek/Class/Characteristic.pdf。

36.4.1 Cayley-Hamilton 定理

根据 Cayley-Hamilton 定理，我们得到矩阵 A 的如下结果：

$$A^n + p_1 A^{n-1} + \ldots + p_{n-1} A + p_n I = 0$$

其中 A^n 是 A 的 n 次方。例如，A^3 是 AAA，矩阵 A 与其自身相乘三次。p 是要确定的系数，I 是单位矩阵，0 是零矩阵。

将前面的方程式乘以 A^{-1}，除以 $-p_n$，重新排列各项，然后得到一个逆方程式，如下所示：

$$\frac{1}{-p_n}\left[A^{n-1}+p_1A^{n-2}+\cdots+p_{n-2}A+p_{n-1}I\right]=A^{-1}$$

因此，要找到矩阵 A 的逆，我们需要进行多次矩阵乘法，并且需要一种方法来找到 p。

对于 4×4 矩阵 A，我们有以下结果：

$$\frac{1}{-p_4}\left[A^3+p_1A^2+p_2A+p_3I\right]=A^{-1}$$

36.4.2 Leverrier 算法

我们使用 Leverrier 算法来计算 p 系数，该算法也在 http://www.mcs.csueastbay.edu/~malek/Class/Characteristic.pdf 中进行了介绍。首先，我们找到矩阵的迹线，即从左上角到右下角的对角线上的元素之和。我们将 s_n 称为矩阵 A^n 的轨迹。

对于 4×4 矩阵 A，我们计算 A 的幂矩阵的迹线，如下所示：

s_1 代表 A
s_2 代表 AA
s_3 代表 AAA
s_4 代表 $AAAA$

Leverrier 给了我们以下信息：

$$p_1 = -s_1$$

$$p_2 = -\frac{1}{2}(p_1s_1+s_2)$$

$$p_3 = -\frac{1}{3}(p_2s_1+p_1s_2+s_3)$$

$$p_4 = -\frac{1}{4}(p_3s_1+p_2s_2+p_1s_3+s_4)$$

很简单吧！当然，需要使用一些复杂的矩阵乘法来获得迹线。

36.4.3 代码

在 inverse4x4 函数中，有一个单独的 .data 段，其中放置了单位矩阵和一些稍后将使用的变量。首先，我们计算幂矩阵并将其存储在 matrix2、matrix3 和 matrix4 中。目前还不会使用 matrix1。然后为每个矩阵调用函数 vtrace 来计算迹线。在 vtrace 函数中，首先在 ymm 寄存器(ymm0、ymm1、ymm2、ymm3)中构建矩阵，每个寄存器都包含一行。然后，我们使用指令 vblendpd，该指令具有四个操作数：两个源操作数、一个目标操作数和一个控制掩码。我们要提取第 2、3 和 4 行中的对角线元素，并将它们作为打包值放在 ymm0 中的位置索引 1、2 和 3 处。在位置 0 处，我们保留 ymm0 的迹元素(Trace element)[1]。

掩码确定从源操作数中选择哪些打包值。如果掩码中的 1 表示此位置，从第二个源操作数中选择值。如果掩码中的 0 表示此位置，从第一个源操作数中选择值。原理图参见图 36-1，但注意，在图中我们以与位掩码索引对应的方式在寄存器中显示值。在调试器中，你将看到 ymm0 中的位置是 $a1$、$a0$、$a3$、$a2$。

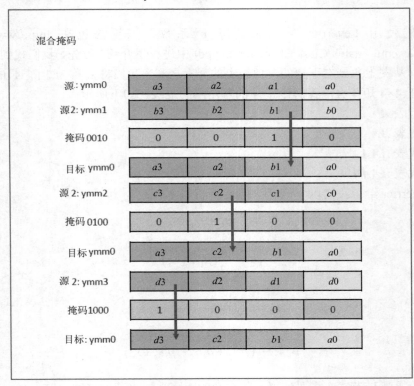

图 36-1　混合掩码

在第一次迹计算中，混合之后 ymm0 寄存器包含迹元素 2、13、29、47。你可以使用 SASM 进行检查。请勿被 ymm0 值的显示顺序所欺骗：13、2、47、29。现

1　译者注：迹或迹线(trace)是指矩阵主对角线(从左上方至右下方的对角线)上所有元素之和。

第 36 章 AVX 矩阵运算

在，必须对这些值进行求和。可以通过提取和简单的加法来轻松完成此操作，但是为了便于演示，将使用 AVX 指令。应用水平加法指令 vhaddpd。然后，ymm0 包含 15、15、76、76，它们是两个较低值的和与两个较高值的和。然后，使用掩码 00100111 执行一个组合 vpermpd。每个 2-bit 值在源操作数中选择一个值；有关说明请参见图 36-2。现在 ymm0 的低位部分 xmm0 包含两个值，因此必须将它们相加以获得迹。使用 haddpd 在 xmm0 上执行水平添加。将迹存储在 xmm8、xmm9、xmm10 和 xmm11 中，以备后用。

你可能会觉得这样做有点过头了。这样做只是为了展示几个 AVX 指令和如何使用掩码。

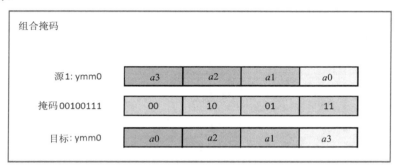

图 36-2　组合掩码

当有了所有的迹时，就可以计算 p 系数。看看我们如何通过应用减号掩码和指令 vxorpd 来更改值的符号。我们使用 vfmadd213sd 和 vfmadd231sd 在一条指令中执行加法和乘法。指令 vfmadd213sd 表示将第一个和第二个操作数相乘，再加上第三个操作数，然后将结果放入第一个操作数。指令 vfmadd231sd 表示将第二个和第三个操作数相乘，将第一个操作数相加，然后将结果放入第一个操作数中。英特尔手册中包含类似说明。

当拥有所有系数时，根据前面的公式将 matrix、matrix2、matrix3 和 matrixI 与系数进行标量乘积。与矩阵相乘的结果放在 matrix1 中。我们不再需要 matrix4，因此为了节省内存，可将用于求逆的空间用作临时内存，而不是使用 matrix4。

必须除以系数 $p4$，因此需要检查 $p4$ 是否为非零。这种情况下，可在之前计算 $p4$ 后完成这个简单操作，但这里想展示如何使用 mxcsr 寄存器。在 mxcsr 中设置除以零掩码位，并使用指令 vdivsd 进行除法运算。如果在除法运算后，mxcsr 寄存器中的第三位(索引 2)被设置，那么可得到一个除以零运算，且矩阵是单数的，不能求逆。在 and 指令中，使用了十进制 4，即二进制的 0000 0100，因此确实在检查第三位。如果得到一个除以零运算，我们在 rax 中使用 1 准备退出(exit)，向调用者发出错误信号。

当矩阵是单数时，程序不会崩溃，因为 mxcsr 寄存器中的除以零运算在默认情况下是经过掩码处理的。完成代码分析后，注释掉检查除以零运算的部分，看看会发生什么。

如果 p_4 不为零，将四个矩阵相加，然后将结果与 $-1/p_4$ 相乘。我们在同一个循环中做加法和乘法。当一切顺利时，可得到相逆的结果，此后 rax 为 0，并返回到调用方。

输出如图 36-3 所示。

```
jo@UbuntuDesktop:~/Desktop/linux64/gcc/44 avx_matrix$ make
nasm -f elf64 -g -F dwarf matrix4x4.asm -l matrix4x4.lst
gcc -o matrix4x4 matrix4x4.o -no-pie
jo@UbuntuDesktop:~/Desktop/linux64/gcc/44 avx_matrix$ ./matrix4x4

4x4 DOUBLE PRECISION FLOATING POINT MATRICES
This is matrixA:
1.000000        3.000000        5.000000        7.000000
9.000000        11.000000       13.000000       15.000000
17.000000       19.000000       21.000000       23.000000
25.000000       27.000000       29.000000       31.000000

This is matrixB:
2.000000        4.000000        6.000000        8.000000
10.000000       12.000000       14.000000       16.000000
18.000000       20.000000       22.000000       24.000000
26.000000       28.000000       30.000000       32.000000

This is matrixA x matrixB:
304.000000      336.000000      368.000000      400.000000
752.000000      848.000000      944.000000      1040.000000
1200.000000     1360.000000     1520.000000     1680.000000
1648.000000     1872.000000     2096.000000     2320.000000

This is matrixC:
2.000000        11.000000       21.000000       37.000000
3.000000        13.000000       23.000000       41.000000
5.000000        17.000000       29.000000       43.000000
7.000000        19.000000       31.000000       47.000000

This is the inverse of matrixC:
1.000000        -1.000000       -1.000000       1.000000
-2.000000       1.833333        0.944444        -0.888889
1.000000        -1.100000       -0.066667       0.233333
0.000000        0.133333        -0.188889       0.077778

Proof: matrixC x inverse =
1.000000        0.000000        0.000000        0.000000
0.000000        1.000000        0.000000        0.000000
-0.000000       -0.000000       1.000000        -0.000000
0.000000        0.000000        -0.000000       1.000000

This is matrixS:
1.000000        2.000000        3.000000        4.000000
5.000000        6.000000        7.000000        8.000000
9.000000        10.000000       11.000000       12.000000
13.000000       14.000000       15.000000       16.000000

This matrix is singular!
```

图 36-3　matrix4x4.asm 的输出

36.5 小结

本章内容：
- AVX 矩阵运算
- 具有三个操作数的 AVX 指令
- Leverrier 算法
- 使用掩码进行混合和组合

第 37 章

矩阵转置

下面做最后一个有用的矩阵运算：转置。我们已经编写了两个版本，一个使用解包，一个使用重排。

37.1 转置代码示例

代码清单 37-1 展示了代码。

代码清单 37-1：transpose4x4.asm

```
; transpose4x4.asm
extern printf

section .data
        fmt0    db      "4x4 DOUBLE PRECISION FLOATING POINT MATRIX
                        TRANSPOSE",10,0
        fmt1    db      10,"This is the matrix:",10,0
        fmt2    db      10,"This is the transpose (unpack):",10,0
        fmt3    db      10,"This is the transpose (shuffle):",10,0

        align   32
        matrix  dq      1., 2., 3., 4.
                dq      5., 6., 7., 8.
                dq      9., 10., 11., 12.
                dq      13., 14., 15., 16.
section .bss
        alignb    32
        transpose resd 16
```

```nasm
section .text
        global main
main:
push rbp
mov rbp,rsp

; 打印标题
        mov rdi, fmt1
        call printf

; 打印矩阵
        mov rdi,fmt1
        call printf
        mov rsi,matrix
        call printm4x4

; 计算转置解包
        mov rdi, matrix
        mov rsi, transpose
        call transpose_unpack_4x4

;打印结果
        mov rdi, fmt2
        xor   rax,rax
        call printf
        mov rsi, transpose
        call printm4x4

; 计算转置重排
        mov rdi, matrix
        mov rsi, transpose
        call transpose_shuffle_4x4

;打印结果
        mov rdi, fmt3
        xor   rax,rax
        call printf
        mov rsi, transpose
        call printm4x4
leave
ret
```

第 37 章 ■ 矩阵转置

```
;----------------------------------------------------
transpose_unpack_4x4:
push rbp
mov rbp,rsp
;将矩阵加载到寄存器中
        vmovapd    ymm0,[rdi]           ; 1 2 3 4
        vmovapd    ymm1,[rdi+32]        ; 5 6 7 8
        vmovapd    ymm2,[rdi+64]        ; 9 10 11 12
        vmovapd    ymm3,[rdi+96]        ; 13 14 15 16
;解包
        vunpcklpd  ymm12,ymm0,ymm1      ; 1 5 3 7
        vunpckhpd  ymm13,ymm0,ymm1      ; 2 6 4 8
        vunpcklpd  ymm14,ymm2,ymm3      ; 9 13 11 15
        vunpckhpd  ymm15,ymm2,ymm3      ; 10 14 12 16
;交换
        vperm2f128 ymm0,ymm12,ymm14, 00100000b ; 1 5 9 13
        vperm2f128 ymm1,ymm13,ymm15, 00100000b ; 2 6 10 14
        vperm2f128 ymm2,ymm12,ymm14, 00110001b ; 3 7 11 15
        vperm2f128 ymm3,ymm13,ymm15, 00110001b ; 4 8 12 16
;写入内存
        vmovapd    [rsi], ymm0
        vmovapd    [rsi+32],ymm1
        vmovapd    [rsi+64],ymm2
        vmovapd    [rsi+96],ymm3
leave
ret

;----------------------------------------------------
transpose_shuffle_4x4:
push rbp
mov rbp,rsp
;将矩阵加载到寄存器中
        vmovapd  ymm0,[rdi]       ; 1 2 3 4
        vmovapd  ymm1,[rdi+32]    ; 5 6 7 8
        vmovapd  ymm2,[rdi+64]    ; 9 10 11 12
        vmovapd  ymm3,[rdi+96]    ; 13 14 15 16
;重排
        vshufpd  ymm12,ymm0,ymm1, 0000b ; 1 5 3 7
        vshufpd  ymm13,ymm0,ymm1, 1111b ; 2 6 4 8
        vshufpd  ymm14,ymm2,ymm3, 0000b ; 9 13 11 15
```

293

```
        vshufpd    ymm15,ymm2,ymm3, 1111b  ; 10 14 12 16
    ;交换
        vperm2f128  ymm0,ymm12,ymm14, 00100000b ; 1 5 9 13
        vperm2f128  ymm1,ymm13,ymm15, 00100000b ; 2 6 10 14
        vperm2f128  ymm2,ymm12,ymm14, 00110001b ; 3 7 11 15
        vperm2f128  ymm3,ymm13,ymm15, 00110001b ; 4 8 12 16
    ;写入内存
        vmovapd [rsi], ymm0
        vmovapd [rsi+32],ymm1
        vmovapd [rsi+64],ymm2
        vmovapd [rsi+96],ymm3
    leave
    ret
    ;--------------------------------------------------------
    printm4x4:
    section .data
        .fmt  db "%.f",9,"%.f",9,"%.f",9,"%.f",10,0
section .text
    push rbp
    mov rbp,rsp
    push  rbx          ;被调用者保留
    push  r15          ;被调用者保留
        mov rdi,.fmt
        mov rcx,4
        xor  rbx,rbx        ;行计数器
    .loop:
        movsd  xmm0, [rsi+rbx]
        movsd  xmm1, [rsi+rbx+8]
        movsd  xmm2, [rsi+rbx+16]
        movsd  xmm3, [rsi+rbx+24]
        mov         rax,4       ;四个浮点数
          push rcx              ;调用者保留
          push rsi              ;调用者保留
          push rdi              ;调用者保留
          ;如果需要,对齐堆栈
          xor   r15,r15
          test  rsp,0fh         ;最后一个字节是8（没有对齐）？
          setnz       r15b      ;如果没有对齐,则设置
          shl   r15,3           ;乘以8
```

```
            sub   rsp,r15        ;减去 0 或 8
            call  printf
            add   rsp,r15        ;加 0 或 8
            pop   rdi
            pop   rsi
            pop   rcx
            add   rbx,32         ;下一行
            loop  .loop
    pop r15
    pop rbx
    leave
    ret
```

输出如图 37-1 所示。

```
jo@ubuntu18:~/Desktop/Book/45 avx_transpose$ ./transpose4x4
This is the matrix:
This is the matrix:
1      2      3      4
5      6      7      8
9      10     11     12
13     14     15     16

This is the transpose (unpack):
1      5      9      13
2      6      10     14
3      7      11     15
4      8      12     16

This is the transpose (shuffle):
1      5      9      13
2      6      10     14
3      7      11     15
4      8      12     16
jo@ubuntu18:~/Desktop/Book/45 avx_transpose$
```

图 37-1   transpose4x4.asm

37.2   解包版本

首先是关于小端格式和打包的 ymm 值的说法。在本例中，当我们拥有第 1、2、3、4 行时，小端格式将是 4、3、2、1。但是，因为在我们的示例中 ymm 存储打包的值，所以 SASM 中的 ymm 看起来是这样的：2、1、4、3。你可以使用调试器验证这一点。在调试程序时，这可能造成混淆。接下来将使用 4、3、2、1 的小端格式，而不会使用 2、1、4、3 格式。

牢记前面的说明,将矩阵加载到 ymm 寄存器中时,这些寄存器具有以下布局(括

号中的示例值):

ymm0	high qword2 (4)	low qword2 (3)	high qword1 (2)	low qword1 (1)
ymm1	high qword4 (8)	low qword4 (7)	high qword3 (6)	low qword3 (5)
...				

以下是 vunpcklpd 指令：

```
vunpcklpd ymm12,ymm0,ymm1
```

从操作数 2 和 3 中获取第一个低位四字，将其存储在操作数 1 中，然后以类似的方式获取第二个低位四字以生成以下结果：

ymm12	low qword4 (7)	low qword2 (3)	low qword3 (5)	low qword1 (1)

类似地，指令 vunpckhpd 从操作数 2 和 3 中提取高位数，然后以类似方式将它们存储在操作数 1 中。

```
vunpckhpd ymm13,ymm0,ymm1
```

ymm13	high qword4 (8)	high qword2 (4)	high qword3 (6)	high qword1 (2)

这种解包的目的是将列对更改为行对。例如，$[^1_5]$ 变为 [1 5]。

解包后，ymm 寄存器以小端格式显示如下：

ymm12	7	3	5	1
ymm13	8	4	6	2
ymm14	15	11	13	9
ymm15	16	12	14	10

使用人类可读格式，而不是小端字节序格式，内容如下：

1	5	3	7
2	6	4	8
9	13	11	15
10	14	12	16

现在，我们必须在行之间排列值，以正确的顺序获取值。我们需要获得小端格式的以下内容：

第 37 章 ■ 矩阵转置

13	9	5	1
14	10	6	2
15	11	7	3
16	12	8	4

你可能注意到，ymm12 和 ymm13 的两个较低值位于正确位置。类似地，ymm14 和 ymm15 的两个上限值位于正确的位置。

我们必须将 ymm14 的两个较低值移至 ymm12 的较高值，并将 ymm15 的两个较低值移至 ymm13 的较高值。

ymm12 的两个较高值必须移到 ymm14 的较低值，我们希望 ymm13 的两个较高值进入 ymm15 的低位。

这样做的操作称为置换。使用 vperm2f128，我们可以置换成对的两个值(128 位)。我们使用掩码来控制排列：例如，掩码 00110001 表示从低位开始。请记住，在以下说明中，索引从 0 开始。

- **01**：从源 1 取出 128 字节的高位字段，并将其放在目标位置 0
- **00**：具有特殊含义；请参阅以下说明。
- **11**：从源 2 取出 128 字节的高位字段，并将其放在目标位置 128。
- **00**：具有特殊含义；请参阅以下说明。

这里我们再次使用小端格式(4、3、2、1)，并且不考虑这些值在 ymm 寄存器中的存储顺序。

因此，实际上，两个源的两个 128 位字段是按顺序编号的。

- 源 1 低位字段 = 00
- 源 1 高位字段 = 01
- 源 2 低位字段 = 10
- 源 2 高位字段 = 11

特殊含义(Special meaning)表示，如果在掩码中设置了第三位(索引 3)，则目标低位字段将被归零；如果在掩码中设置了第七位(索引 7)，则目标高位字段将被归零。

这里不使用第二、第三、第六和第七位。大多数情况下，你可按如下方式读取诸如 00110001 的掩码：00110001。

以下是程序的代码：

```
vperm2f128 ymm0, ymm12, ymm14, 00100000b
```

这里我们得到 00**100000**。
- 较低的 00 表示将 ymm12 的低字段(5, 1)放入 ymm0 的低位字段。
- 较高的 10 表示将 ymm14 的低字段(13, 9)放入 ymm0 的高位字段。

ymm12	7	3	5	1
ymm14	15	11	13	9
ymm0	13	9	5	1

现在 ymm0 包含一个已完成的行。然后是下一行。

```
vperm2f128 ymm1, ymm13, ymm15, 00100000b
```

这里我们得到 00**100000**。
- 较低的 00 表示将 ymm13 的低字段(6, 2)放入 ymm1 的低位字段。
- 较高的 10 表示将 ymm15 的低字段(14, 10)放入 ymm1 的高位字段。

ymm13	8	4	6	2
ymm15	16	12	14	10
ymm1	14	10	6	2

现在 ymm1 包含一个已完成的行。然后是下一行。

```
vperm2f128 ymm2, ymm12, ymm14, 00110001b
```

这里我们得到 00**100001**。
- 较低的 01 表示将 ymm13 高字段(7, 3)放入 ymm2 的低位字段中。
- 较高的 11 表示将 ymm15 的高字段 (15, 11)放入 ymm2 的高位字段中。

ymm12	7	3	5	1
ymm14	15	11	13	9
ymm2	15	11	7	3

现在 ymm2 包含一个已完成的行。最后一行。

```
vperm2f128 ymm3, ymm13, ymm15, 00110001b
```

这里我们得到 00**100001**。
- 较低的 01 表示将 ymm13 高字段(8, 4)放入 ymm3 的低位字段中。
- 较高的 11 表示将 ymm15 的高字段(16, 12)放入 ymm3 的高位字段中。

ymm13	8	4	6	2
ymm15	16	12	14	10
ymm3	16	12	8	4

我们已经完成了排列。剩下的就是将 ymm 寄存器中的行复制到内存中的正确顺序。

37.3 重排版本

在第 33 章中，我们使用了一个名为 pshufd 的随机重排指令。在这里，我们使用 vshufpd 指令，该指令也使用掩码来控制随机重排。不要觉得困惑，指令 pshufd 使用 8 位掩码。我们将在此处使用的掩码仅为 4 位。

同样，我们使用小端格式(记住 4、3、2、1)，并且不关心打包值如何存储在 ymm 寄存器中。那是处理器的事情。

请参阅下表和本说明后面的示例。掩码中的两个低位控制哪些打包值进入目的地的两个低位；掩码中的两个高位控制哪些打包值进入目的地的两个高位。位 0 和 2 指定从源 1 取哪个值，位 1 和 3 指定从源 2 取哪个值。

从源 2 的两个高位值选择	从源 1 的两个高位值选择	从源 2 的两个低位值选择	从源 1 的两个低位值选择
0=源 2 的低位值	0=源 1 的低位值	0=源 2 的低位值	0=源 1 的低位值
1=源 2 的高位值	1=源 1 的高位值	1=源 2 的高位值	1=源 1 的高位值

每个源中的两个较低值永远不会在目标位置的较高位置结束，并且每个源中的两个较高位值永远不会在目标位置的较低位置结束。参见图 37-2 的几个样例掩码示意图。

```
                重排掩码

    源1: ymm0      | a3 | a2 | a1 | a0 |
    源2: ymm1      | b3 | b2 | b1 | b0 |
    掩码 0000      | 0  | 0  | 0  | 0  |
    目的: ymm3     | b2 | a2 | b0 | a0 |

    源1: ymm0      | a3 | a2 | a1 | a0 |
    源2: ymm1      | b3 | b2 | b1 | b0 |
    掩码 1111      | 1  | 1  | 1  | 1  |
    目标: ymm3     | b3 | a3 | b1 | a1 |

    源1: ymm0      | a3 | a2 | a1 | a0 |
    源2: ymm1      | b3 | b2 | b1 | b0 |
    掩码 0110      | 0  | 1  | 1  | 0  |
                   | b2 | a3 | b1 | a0 |

    源1: ymm0      | a3 | a2 | a1 | a0 |
    源2: ymm1      | b3 | b2 | b1 | b0 |
    掩码 0011      | 0  | 0  | 1  | 1  |
                   | b2 | a2 | b1 | a1 |
```

图 37-2 重排掩码示例

程序中的代码如下：

```
vshufpd ymm12,ymm0,ymm1, 0000b
```

ymm0	4	3	2	1
ymm1	8	7	6	5
ymm12	低上位 ymm1 7	低上位 ymm0 3	低下位 ymm1 5	低下位 ymm0 1

```
vshufpd ymm13,ymm0,ymm1, 1111b
```

ymm0	4	3	2	1
ymm1	8	7	6	5
ymm13	高上位 ymm1	高上位 ymm0	高下位 ymm1	高下位 ymm0
	8	4	6	2

```
vshufpd ymm14,ymm2,ymm3, 0000b
```

ymm2	12	11	10	9
ymm3	16	15	14	13
ymm14	低上位 ymm3	低上位 ymm2	低下位 ymm3	低下位 ymm2
	15	11	13	9

最后一个例子：

```
vshufpd ymm15,ymm2,ymm3, 1111b
```

ymm2	12	11	10	9
ymm3	16	15	14	13
ymm15	高上位 ymm3	高上位 ymm2	高下位 ymm3	高下位 ymm2
	16	12	14	10

应用重排掩码后，ymm 寄存器中有八对值。我们选择寄存器，以便获得与解包版本相同的中间结果。现在需要在正确的位置重新排列成对，以形成转置。我们用与解包部分完全相同的方法来实现这一点，即使用 vperm2f128 排列 128 位的字段(块)。

37.4 小结

本章内容：
- 有两种转置矩阵的方法
- 如何使用重排、解包和排列指令
- 重排、解包和排列有不同的掩码

第 38 章

性能调优

相信你也认为很多 AVX 指令都很不直观，特别是不同的掩码布局使得代码很难阅读和理解。此外，位掩码有时是用十六进制表示法编写的，因此必须首先将它们转换为二进制表示法，这样才能查看它们做了什么。

在本章中，将演示使用 AVX 指令可以显著提高性能，并且在许多情况下使用 AVX 所付出的努力是有回报的。可以在 https://www.intel.com/content/dam/www/public/us/en/documents/white-papers/ia-32-ia-64-benchmarkcode-execution-paper.pdf 找到有关基准测试代码的白皮书。

在示例中，将使用该白皮书中介绍的测量方法。

38.1 转置计算性能

在代码清单 38-1 所示的示例代码中，我们有两种计算转置矩阵的方法，一种使用"经典"汇编程序指令，另一种使用 AVX 指令。我们添加了代码来测量两种算法的执行时间。

代码清单 38-1：transpose.asm

```
; transpose.asm
extern printf

section .data
    fmt0    db    "4x4 DOUBLE PRECISION FLOATING POINT MATRIX
                   TRANSPOSE",10,0
    fmt1    db    10,"This is the matrix:",10,0
    fmt2    db    10,"This is the transpose (sequential version): ",10,0
    fmt3    db    10,"This is the transpose (AVX version): ",10,0
```

```nasm
        fmt4    db      10,"Number of loops: %d",10,0
        fmt5    db      "Sequential version elapsed cycles: %d",10,0
        fmt6    db      "AVX Shuffle version elapsed cycles: %d",10,0

        align 32
        matrix  dq 1., 2., 3., 4.
                dq 5., 6., 7., 8.
                dq 9., 10., 11., 12.
                dq 13., 14., 15., 16.

        loops   dq 10000
section .bss
        alignb          32
        transpose       resq 16

        bahi_cy         resq 1          ;avx 版本的计时器
        balo_cy         resq 1
        eahi_cy         resq 1
        ealo_cy         resq 1

        bshi_cy         resq 1          ;顺序版本的定时器
        bslo_cy         resq 1
        eshi_cy         resq 1
        eslo_cy         resq 1

section .text
        global main
main:
push rbp
mov rbp,rsp
; 打印标题
        mov rdi, fmt0
        call printf
; 打印矩阵
        mov rdi,fmt1
        call printf
        mov rsi,matrix
        call printm4x4
; SEQUENTIAL VERSION(顺序版本)
; 计算转置
```

```
        mov rdi, matrix
        mov rsi, transpose
        mov rdx, [loops]

;开始测量周期
        cpuid
        rdtsc
        mov [bshi_cy],edx
        mov [bslo_cy],eax

        call seq_transpose

;停止测量周期
        rdtscp
        mov [eshi_cy],edx
        mov [eslo_cy],eax
        cpuid

;打印结果
        mov rdi,fmt2
        call printf
        mov rsi,transpose
        call printm4x4

; AVX VERSION(AVX 版本)
; 计算转置
        mov rdi, matrix
        mov rsi, transpose
        mov rdx, [loops]
;开始测量周期
        cpuid
        rdtsc
        mov [bahi_cy],edx
        mov [balo_cy],eax

        call AVX_transpose

;停止测量周期
        rdtscp
        mov [eahi_cy],edx
        mov [ealo_cy],eax
        cpuid
```

```
;打印结果
        mov rdi,fmt3
        call printf
        mov rsi,transpose
        call printm4x4

;打印循环
        mov rdi,fmt4
        mov rsi,[loops]
        call printf
;打印循环
;循环顺序版本
        mov rdx,[eslo_cy]
        mov rsi,[eshi_cy]
        shl  rsi,32
        or   rsi,rdx        ;rsi 包含结束时间

        mov r8,[bslo_cy]
        mov r9,[bshi_cy]
        shl  r9,32
        or   r9,r8          ;r9 包含开始时间

        sub  rsi,r9         ;rsi 包含经过的时间
;打印时间结果
        mov rdi,fmt5
        call printf

;循环 AVX 混合版本
        mov rdx,[ealo_cy]
        mov rsi,[eahi_cy]
        shl  rsi,32
        or   rsi,rdx        ;rsi 包含结束时间
        mov r8,[balo_cy]
        mov r9,[bahi_cy]
        shl  r9,32
        or   r9,r8          ;r9 包含开始时间

        sub rsi,r9          ;rsi 包含经过的时间
;打印时间结果
        mov rdi,fmt6
        call printf
```

```
            leave
            ret
            ;--------------------------------------------------
            seq_transpose:
            push rbp
            mov rbp,rsp
                .loopx:              ; 循环次数
                    pxor    xmm0,xmm0
                    xor     r10,r10
                    xor     rax,rax
                    mov     r12,4
                    .loopo:
                            push rcx
                            mov   r13,4
                            .loopi:
                                    movsd xmm0, [rdi+r10]
                            movsd [rsi+rax], xmm0
                            add         r10,8
                            add         rax,32
                            dec         r13
                            jnz         .loopi
                            add         rax,8
                            xor         rax,10000000b     ;rax - 128
                            inc         rbx
                            dec         r12
                    jnz .loopo
                    dec rdx
            jnz .loopx
            leave
            ret
            ;--------------------------------------------------
            AVX_transpose:
            push rbp
            mov rbp,rsp
                .loopx:              ; 循环次数
                ;将矩阵加载到寄存器中
                        vmovapd     ymm0,[rdi]           ; 1 2 3 4
                        vmovapd     ymm1,[rdi+32]        ; 5 6 7 8
                        vmovapd     ymm2,[rdi+64]        ; 9 10 11 12
```

```
            vmovapd    ymm3,[rdi+96]       ; 13 14 15 16
;重排
            vshufpd    ymm12,ymm0,ymm1, 0000b  ; 1 5 3 7
            vshufpd    ymm13,ymm0,ymm1, 1111b  ; 2 6 4 8
            vshufpd    ymm14,ymm2,ymm3, 0000b  ; 9 13 11 15
            vshufpd    ymm15,ymm2,ymm3, 1111b  ; 10 14 12 16
;交换
            vperm2f128 ymm0,ymm12,ymm14, 00100000b ; 1 5 9 13
            vperm2f128 ymm1,ymm13,ymm15, 00100000b ; 2 6 10 14
            vperm2f128 ymm2,ymm12,ymm14, 00110001b ; 3 7 11 15
            vperm2f128 ymm3,ymm13,ymm15, 00110001b ; 4 8 12 16
;写入内存
            vmovapd    [rsi], ymm0
            vmovapd    [rsi+32],ymm1
            vmovapd    [rsi+64],ymm2
            vmovapd    [rsi+96],ymm3
            dec        rdx
            jnz        .loopx
leave
ret
;-------------------------------------------------------------
printm4x4:
section .data
       .fmt   db "%f",9,"%f",9, "%f",9,"%f",10,0
section .text
push   rbp
mov    rbp,rsp
       push rbx         ;被调用者保留
       push r15         ;被调用者保留
       mov rdi,.fmt
       mov rcx,4
       xor  rbx,rbx     ;行计数器
.loop:
       movsd   xmm0, [rsi+rbx]
       movsd   xmm1, [rsi+rbx+8]
       movsd   xmm2, [rsi+rbx+16]
       movsd   xmm3, [rsi+rbx+24]
       mov         rax,4        ; 四个浮点数
         push        rcx        ;调用者保留
```

```
            push      rsi            ;调用者保留
            push      rdi            ;调用者保留
            ;align stack if needed
            xor       r15,r15
            test      rsp,0fh        ;最后一个字节是 8 (没有对齐)?
            setnz     r15b           ;如果没有对齐,则设置
            shl       r15,3          ;乘以 8
            sub       rsp,r15        ;减去 0 或 8
            call      printf
            add       rsp,r15        ;加 0 或 8
            pop       rdi
            pop       rsi
            pop       rcx
            add       rbx,32         ;下一行
            loop      .loop
pop r15
pop rbx
leave
ret
```

我们在调用转置函数之前开始进程计时。现代处理器支持乱序的执行代码,这在开始计时之前或停止计时之后可能导致这样的错误时刻执行指令。为了避免这种情况,我们需要使用"序列化"指令,这些指令可以确保时序指令只测量我们想要测量的内容。有关更详细的说明,请参见前面提到的白皮书。cpuid 是可用于序列化的一种此类指令。我们在使用 rdtsc 启动计时器之前执行 cpuid。使用 rdtsc 在 eax 寄存器中写入开始时间戳计数器"低周期(low cycles)",在 edx 中写入"高周期(high cycles)",这些值都存储在内存中。出于历史原因,rdtsc 指令使用两个寄存器:对于 32 位的处理器,一个寄存器太小而无法容纳定时器计数。因此,一个 32 位寄存器用于定时器计数器值的低位部分,另一个寄存器用于高位部分。记录开始计时器的计数器的值后,执行要测量的代码,并使用 rdtscp 指令停止测量。最后的"高周期"和"低周期"计数器再次存储在内存中,cpuid 将再次执行,以确保处理器不会延迟指令的执行。

我们使用的是 64 位处理器环境,因此将较高的时间戳值左移 32,然后将较高的时间戳值与较低的时间戳值进行异或运算,以获得 64 位寄存器中的完整时间戳。起始计数器值和终止计数器值之间的差给出了使用的周期数。

函数 seq_transpose 使用"经典"指令,并且函数 AVX_transpose 是上一章中的 transpose_shuffle4x4 函数。这些函数按照变量循环中的指定执行多次。

输出如图 38-1 所示。

```
jo@UbuntuDesktop:~/Desktop/linux64/gcc/46 performance1$ make
nasm -f elf64 -g -F dwarf transpose.asm -l transpose.lst
gcc -o transpose transpose.o -no-pie
jo@UbuntuDesktop:~/Desktop/linux64/gcc/46 performance1$ ./transpose
4x4 DOUBLE PRECISION FLOATING POINT MATRIX TRANSPOSE

This is the matrix:
1.000000        2.000000        3.000000        4.000000
5.000000        6.000000        7.000000        8.000000
9.000000        10.000000       11.000000       12.000000
13.000000       14.000000       15.000000       16.000000

This is the transpose (sequential version):
1.000000        5.000000        9.000000        13.000000
2.000000        6.000000        10.000000       14.000000
3.000000        7.000000        11.000000       15.000000
4.000000        8.000000        12.000000       16.000000

This is the transpose (AVX version):
1.000000        5.000000        9.000000        13.000000
2.000000        6.000000        10.000000       14.000000
3.000000        7.000000        11.000000       15.000000
4.000000        8.000000        12.000000       16.000000

Number of loops: 10000
Sequential version elapsed cycles: 132687
AVX Shuffle version elapsed cycles: 12466
jo@UbuntuDesktop:~/Desktop/linux64/gcc/46 performance1$
```

图 38-1 transpose.asm 的输出

可以看到，使用 AVX 指令可以显著加快处理速度。

英特尔有一个专门用于代码优化的文档：https://software.intel.com/sites/default/files/managed/9e/bc/64-ia-32-architectures-optimizationmanual.pdf。

这本手册有很多关于提高汇编代码性能的有趣信息。在 handling port 5 pressure(在第 14 章中介绍)部分，你将发现 8×8 矩阵的转置算法的几个版本以及不同指令对性能的影响。在上一章中，我们演示了两种转置矩阵的方法：使用解包和重排。英特尔手册对这个主题的细节有更深入的介绍。如果性能对你很重要，那么你会在这里发现很多宝藏。

38.2 迹计算性能

下面是一个说明 AVX 指令并不总是比"经典"汇编指令快的示例。本示例计算 8×8 矩阵的迹(trace)：

```
; trace.asm
extern printf
```

```
section .data
    fmt0 db "8x8 SINGLE PRECISION FLOATING POINT MATRIX TRACE",10,0
    fmt1 db 10,"This is the matrix:",10,0
    fmt2 db 10,"This is the trace (sequential version): %f",10,0
    fmt5 db "This is the trace (AVX blend version): %f",10,0
    fmt6 db 10,"This is the tranpose: ",10,0
    fmt30 db "Sequential version elapsed cycles: %u",10,0
    fmt31 db "AVX blend version elapsed cycles: %d",10,10,0
    fmt4 db 10,"Number of loops: %d",10,0

    align 32
    matrix dd 1., 2., 3., 4., 5., 6., 7., 8.
           dd 9., 10., 11., 12., 13., 14., 15., 16.
           dd 17., 18., 19., 20., 21., 22., 23., 24.
           dd 25., 26., 27., 28., 29., 30., 31., 32.
           dd 33., 34., 35., 36., 37., 38., 39., 40.
           dd 41., 42., 43., 44., 45., 46., 47., 48.
           dd 49., 50., 51., 52., 53., 54., 55., 56.
           dd 57., 58., 59., 60., 61., 62., 63., 64.
    loops  dq 1000
    permps dd 0,1,4,5,2,3,6,7        ;ymm 中置换 sp 值的掩码
section .bss
    alignb   32
    transpose resq 16

    trace     resq 1

    bbhi_cy   resq 1
    bblo_cy   resq 1
    ebhi_cy   resq 1
    eblo_cy   resq 1
    bshi_cy   resq 1
    bslo_cy   resq 1
    eshi_cy   resq 1
    eslo_cy   resq 1

section .text
    global main
main:
push rbp
mov rbp,rsp
```

```nasm
; 打印主题
    mov rdi, fmt0
    call printf
; 打印矩阵
    mov rdi,fmt1
    call printf
    mov rsi,matrix
    call printm8x8
; SEQUENTIAL VERSION(顺序版本)
; 计算迹
    mov rdi, matrix
    mov rsi, [loops]

;开始测量周期
    cpuid
    rdtsc
    mov [bshi_cy],edx
    mov [bslo_cy],eax

    call seq_trace

;停止测量周期
    rdtscp
    mov [eshi_cy],edx
    mov [eslo_cy],eax
    cupid
;打印结果
    mov rdi, fmt2
    mov rax,1
    call printf

; BLEND VERSION(混合版本)
; 计算迹
    mov rdi, matrix
    mov rsi, [loops]

;开始测量周期
    cpuid
    rdtsc
    mov [bbhi_cy],edx
    mov [bblo_cy],eax
```

第 38 章 ■ 性能调优

```
        call blend_trace
;停止测量周期
        rdtscp
        mov [ebhi_cy],edx
        mov [eblo_cy],eax
        cpuid

;打印结果
        mov rdi, fmt5
        mov rax,1
        call printf

;打印循环
        mov rdi,fmt4
        mov rsi,[loops]
        call printf
;打印周期
;周期顺序版本
        mov rdx,[eslo_cy]
        mov rsi,[eshi_cy]
        shl  rsi,32
        or   rsi,rdx

        mov r8,[bslo_cy]
        mov r9,[bshi_cy]
        shl  r9,32
        or   r9,r8

        sub  rsi,r9      ;rsi 包含已用时间
;打印
        mov rdi,fmt30
        call printf

;周期 AVX 混合版本
        mov rdx,[eblo_cy]
        mov rsi,[ebhi_cy]
        shl  rsi,32
        or   rsi,rdx

        mov r8,[bblo_cy]
        mov r9,[bbhi_cy]
```

```
        shl   r9,32
        or    r9,r8

        sub   rsi,r9
        ;打印
        mov rdi,fmt31
        call printf
leave
ret
;--------------------------------------------------------------
seq_trace:
push rbp
mov rbp,rsp
.loop0:
        pxor xmm0,xmm0
        mov rcx,8
        xor   rax,rax
        xor   rbx,rbx
        .loop:
        addss  xmm0, [rdi+rax]
        add rax,36          ;每行32个字节
        loop .loop
        cvtss2sd xmm0,xmm0
        dec   rsi
        jnz   .loop0
leave
ret
;--------------------------------------------------------------
blend_trace:
push rbp
mov rbp,rsp
.loop:
        ;在内存中构建矩阵
            vmovaps    ymm0, [rdi]
            vmovaps    ymm1, [rdi+32]
            vmovaps    ymm2, [rdi+64]
            vmovaps    ymm3, [rdi+96]
            vmovaps    ymm4, [rdi+128]
            vmovaps    ymm5, [rdi+160]
```

```
            vmovaps    ymm6,[rdi+192]
            vmovaps    ymm7,[rdi+224]

            vblendps   ymm0,ymm0,ymm1,00000010b
            vblendps   ymm0,ymm0,ymm2,00000100b
            vblendps   ymm0,ymm0,ymm3,00001000b
            vblendps   ymm0,ymm0,ymm4,00010000b
            vblendps   ymm0,ymm0,ymm5,00100000b
            vblendps   ymm0,ymm0,ymm6,01000000b
            vblendps   ymm0,ymm0,ymm7,10000000b
            vhaddps    ymm0,ymm0,ymm0
            vmovdqu    ymm1,[permps]
            vpermps    ymm0,ymm1,ymm0
            haddps     xmm0,xmm0
            vextractps r8d,xmm0,0
            vextractps r9d,xmm0,1
            vmovd      xmm0,r8d
            vmovd      xmm1,r9d
            vaddss     xmm0,xmm0,xmm1
            dec        rsi
            jnz        .loop
    cvtss2sd xmm0,xmm0
    leave
    ret

    printm8x8:
    section .data
    .fmt db
"%.f,",9,"%.f,",9,"%.f,",9,"%.f,",9,"%.f,",9,"%.f,",9,"%.f,",
    9,"%.f",10,0
    section .text
    push rbp
    mov  rbp,rsp
        push rbx              ;被调用者保留
        mov rdi,.fmt
        mov rcx,8
        xor  rbx,rbx          ;行计数器
        vzeroall
    .loop:
        movss      xmm0, dword[rsi+rbx]
```

```
            cvtss2sd    xmm0,xmm0
        movss           xmm1, [rsi+rbx+4]
            cvtss2sd    xmm1,xmm1
        movss           xmm2, [rsi+rbx+8]
            cvtss2sd    xmm2,xmm2
        movss           xmm3, [rsi+rbx+12]
            cvtss2sd    xmm3,xmm3
        movss           xmm4, [rsi+rbx+16]
            cvtss2sd    xmm4,xmm4
        movss           xmm5, [rsi+rbx+20]
            cvtss2sd    xmm5,xmm5
        movss           xmm6, [rsi+rbx+24]
            cvtss2sd    xmm6,xmm6
        movss           xmm7, [rsi+rbx+28]
            cvtss2sd    xmm7,xmm7
        mov     rax,8           ; 8 个浮点数
        push    rcx             ;调用者保留
        push    rsi             ;调用者保留
        push    rdi             ;调用者保留
          ;如果需要,则对齐栈
        xor     r15,r15
        test    rsp,0fh         ;最后一个字节是 8(没有对齐)?
        setnz   r15b            ;如果没有对齐,则设置
        shl     r15,3           ;乘以 8
        sub     rsp,r15         ;减去 0 或 8
        call    printf
        add     rsp,r15         ;加 0 或 8
        pop     rdi
        pop     rsi
        pop     rcx
        add     rbx,32          ;下一行
        loop    .loop
    pop rbx                     ;被调用者保留
    leave
    ret
```

函数 blend_trace 是我们在第 36 章中使用 AVX 指令在矩阵求逆中使用的迹函数的 4×4 到 8×8 的扩展。函数 seq_trace 依次遍历矩阵,找到迹元素,然后将它们相加。运行此代码时,你将看到 seq_trace 比 blend_trace 快得多。

输出如图 38-2 所示。

```
jo@ubuntu18:~/Desktop/Book/47 performance2$ ./trace
8x8 SINGLE PRECISION FLOATING POINT MATRIX TRACE
This is the matrix:
1,     2,     3,     4,     5,     6,     7,     8
9,     10,    11,    12,    13,    14,    15,    16
17,    18,    19,    20,    21,    22,    23,    24
25,    26,    27,    28,    29,    30,    31,    32
33,    34,    35,    36,    37,    38,    39,    40
41,    42,    43,    44,    45,    46,    47,    48
49,    50,    51,    52,    53,    54,    55,    56
57,    58,    59,    60,    61,    62,    63,    64

This is the trace (sequential version): 260.000000
This is the trace (AVX blend version): 260.000000

Number of loops: 1000
Sequential version elapsed cycles: 48668
AVX blend   version elapsed cycles: 175509
jo@ubuntu18:~/Desktop/Book/47 performance2$
```

图 38-2　trace.asm 的输出

如果想了解有关优化的更多信息，请使用前面提到的英特尔手册。https://www.agner.org 是另一个很好的资源。

38.3　小结

本章内容：

- 测量和计算时间周期
- AVX 可以大大加快处理速度
- AVX 并非适合所有情况

第 39 章

你好，Windows 的世界

在本章和下一章中，我们将开始在 Windows 上编写汇编代码。与 Linux 一样，最好使用 Windows 虚拟机。可以在此处下载 90 天 Windows 10 试用版：https://www.microsoft.com/en-us/evalcenter/evaluate-windows-10-enterprise。安装 Windows 10 的试用版，然后进行更新，这可能需要一些时间。

39.1 入门

微软已经开发了自己的名为 MASM 的汇编器，并且将其包含在 Visual Studio 中。能够使用 Visual Studio 当然是一个优势，因为它是一个全面的开发工具。MASM 中使用的汇编指令与 NASM 中使用的汇编指令相同，但是汇编伪指令有很大不同。配置和学习使用 Visual Studio 有一定的学习曲线，这取决于你以前在 Windows 下的开发经验。

为了缓解不同操作系统带来的影响，在本书中，我们将在 Windows 上使用 NASM 并使用命令行。在前面的章节中我们已经了解了 Linux 上的 NASM，这给了我们一个良好的开端。但是，切换到 MASM 并不是一件容易的事。

如果要开发基于 Windows 的程序，学习使用 Visual Studio 是值得的。甚至可以在互联网上找到如何将 NASM 与 Visual Studio 结合使用的资料。

在互联网上找到适用于 Windows 的 NASM(https://www.nasm.us/pub/nasm/releasebuilds/2.14.03rc2/win64/)并进行安装。确保 Windows 环境路径变量的条目指向安装 NASM 的文件夹。请参见图 39-1。你可以在命令行上使用 nasm -v 验证 NASM 安装。

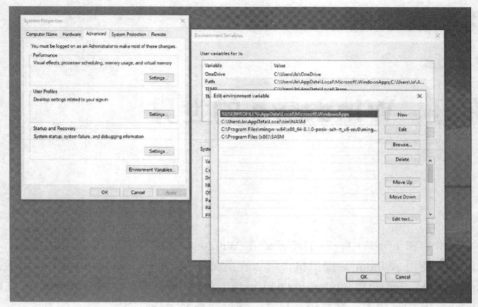

图 39-1　Windows 10 环境路径变量

我们还将使用 MinGW 版本,该版本是移植到 Windows 的一组 Linux 开发工具。MinGW 将允许我们使用在本书前面各章中经常使用的 make 和 GCC 工具。你必须安装的版本是 MinGW-w64。在开始下载和安装之前,如果计划在 Windows 上使用 SASM,需要注意的是 SASM 在其自己的子目录(make 除外)中安装了 NASM 和某些 MinGW-w64 工具。如果你手动安装 SASM 和 MinGW-w64,将最终进行双重安装。在 SASM 设置中,可以将 SASM 配置为使用已安装的 NASM 和 GCC 版本,而不是使用 SASM 随附的旧版本。

当前,你可以在以下位置找到 MinGW-w64 的下载文件:http://mingw-w64.org/doku.php/download。选择 MinGW-W64-builds,下载并安装它,然后在安装窗口中选择 x86_64。

转到 Windows 环境变量,然后将 MinGW-w64 bin 文件夹的路径添加到环境变量路径,如图 39-1 所示。bin 文件夹包含 GCC。更新路径变量后,转到 PowerShell 命令行并输入 gcc -v 以验证安装。

下载 win64 版本的 SASM(https://dman95.github.io/SASM/english.html),如果你希望 SASM 使用新版本的 NASM 和 GCC,则将编译设置修改为新安装的 NASM 和 GCC。不要忘记使用 SASM 条目来更新 Windows 环境路径变量。

如果 Windows 上没有首选的文本编辑器,请安装 Notepad++。它很简单,并为包括汇编在内的多种编程语言提供语法高亮显示功能。而且,你可以轻松地将编码设置为 UTF-8、UTF-16 等。你可以在菜单栏的"语言"下找到汇编语言设置。

令人恼火的是，MinGW-w64 没有 make 命令，仅提供 ming32-make.exe，这是一个很长的命令。要解决此问题，请使用 Notepad++(以管理员身份运行)创建一个 make.bat 文件，其中包含以下行：

```
mingw32-make.exe
```

将文件以 UTF-8 格式保存在 MinGW-w64 bin 文件夹中。

如果你在使用 Windows 时遇到困难，这里有一些帮助提示：

- 要以管理员身份打开应用程序，请右击应用程序图标，然后选择"以管理员身份运行"选项。
- 轻松访问 Windows 的命令行——PowerShell 总是很方便的。要打开它，请在底部任务栏上的搜索框中输入 PowerShell，然后单击"打开"。一个 PowerShell 图标将出现在任务栏上。右击该图标，然后选择"固定到任务栏"。
- 在显示文件或目录图标的窗口中，按住 Shift 键并同时单击鼠标右键，然后在弹出的菜单上选择"在此处打开 PowerShell 窗口"。
- 要显示隐藏的文件和目录，请单击任务栏上的"文件资源管理器"图标。打开"查看"菜单项，然后选择"隐藏的项目"。
- 要查找环境变量，请在任务栏上的搜索框中输入环境变量。

39.2 编写一些代码

现在你可以开始编写代码了。代码清单 39-1 和代码清单 39-2 展示了第一个程序。

代码清单 39-1：hello.asm

```
; hello.asm
extern printf
section .data
      msg db 'Hello, Windows World!',0
      fmt db "Windows 10 says: %s",10,0
section .text
      global main
main:
push rbp
```

```
mov rbp,rsp
    mov rcx, fmt
    mov rdx, msg
    sub rsp,32
    call printf
    add rsp,32
leave
ret
```

代码清单 39-2：makefile

```
hello.exe: hello.obj
    gcc -o hello.exe hello.obj
hello.obj: hello.asm
    nasm -f win64 -g -F cv8 heool.asm -l hello.lst
```

看上去这里没有什么特别的地方。

首先是 sub rsp,32，在 Linux 中，我们曾用它来创建堆栈变量。通过这个指令，我们在调用函数之前在堆栈上创建了影子空间。稍后继续这个话题。在执行 printf 函数后，我们使用 add rsp,32 还原堆栈，在本例中这不是严格必需的，因为堆栈将通过 leave 指令来还原。用来将参数传递给 printf 的寄存器与 Linux 中使用的寄存器不同。这是因为 Windows 中的调用约定与 Linux 中的调用约定不同。Windows 要求使用 Microsoft x64 调用约定，而 Linux 要求使用 System V 应用程序二进制接口(System V Application Binary Interface)，也称为 System V ABI。

可以在此处找到 Microsoft 调用约定的概述：https://docs.microsoft.com/zh-cn/cpp/build/x64-calling-convention?view=vs-2019。该页面经常会有变动；如果找不到，请在 Microsoft 网站上搜索 x64 调用约定。以下是简短的版本：

- 整数参数按 rcx、rdx、r8 和 r9 的顺序传递。
- 如果要传递更多参数，可以将它们压入堆栈。
- 浮点参数在 xmm0-xmm3 寄存器中传递；使用堆栈传递更多参数。
- 寄存器 rcx、rdx、r8、r9 以及另外的 rax、r10、r11、xmm4 和 xmm5 是易失性的，这意味着调用者必须在需要时保存它们。其他寄存器被调用者保存。
- 调用者需要在堆栈上提供一个 32 字节的空间(影子空间)，以便将四个函数参数传递给被调用者，即使被调用者没有接收那么多参数。
- 在 Linux 中，堆栈必须是 16 字节对齐的。

图 39-2 展示了第一个程序的输出。

图 39-2　hello.asm 的输出

39.3　调试

如果启动 GDB 来调试我们的第一个程序，你可能会感到惊讶。你可以执行许多命令，但单步执行代码将不起作用。你将看到以下消息：

```
Single stepping until exit from function main, which has no line number information.
0x0000000000402a60 in printf ()
```

这意味着 GDB 在这里用途有限！SASM 可以解决这个问题。在 makefile 文件中，仍然包含调试标志。也许在 GDB 的未来版本中，这个问题将得到解决。在 makefile 中，我们指定 cv8(Microsoft CodeView 8)作为调试格式。

39.4　syscall

在示例代码中，我们使用 printf 而不是 syscall，就像第一个 Linux 汇编程序那样。原因是：在 Windows 中不使用 syscall。Windows 有 syscall，但它们只供"内部"使用。当你想访问系统资源时，需要使用 Windows API。当然，你可以在 Windows 代码中或在互联网上查找什么是 Windows syscall，但是要知道较新版本的 Windows 可以更改 syscall 的用法；如果你使用它们，可能会破坏你的代码。

39.5　小结

本章内容：
- 如何在 Windows 中安装和使用 NASM、SASM 和 Linux 开发工具
- Windows 中的调用约定与 Linux 中的调用约定不同
- 最好不要使用 syscall

第 40 章

使用 Windows API

Windows 应用程序编程接口(API)是开发人员用来与操作系统进行交互的一组函数。如上一章所述，syscall 不是与操作系统进行通信的可靠方法，但是 Microsoft 提供了许多 API 来完成你可能想到的所有事情。Windows API 是使用 C 编程语言编写的，但是如果我们遵守调用约定，则可以在汇编程序中轻松使用 Windows API。可从以下位置找到 Windows API 的描述：https://docs.microsoft.com/zh-cn/windows/win32/api/。

40.1 控制台输出

代码清单 40-1 展示了 "Hello, World" 程序的一个版本，它使用 Windows API 在屏幕上显示消息。

代码清单 40-1：helloc.asm

```
; helloc.asm
%include "win32n.inc"
    extern WriteFile
    extern WriteConsoleA
    extern GetStdHandle

section .data
    msg db 'Hello, World!!',10,0
    msglen EQU $-msg-1      ; 去掉 NULL

section .bss
    hFile  resq 1           ;文件句柄
```

```
        lpNumberOfBytesWritten resq 1
section .text
        global main
main:
push rbp
mov rbp,rsp

; 获取标准输出的句柄
;HANDLE WINAPI GetStdHandle(
; _In_ DWORD nStdHandle
;);
        mov   rcx, STD_OUTPUT_HANDLE
        sub   rsp,32              ;影子空间
        call  GetStdHandle        ;如果没有成功,则返回 INVALID_HANDLE_VALUE
        add   rsp,32
        mov   qword[hFile],rax    ;将接收到的句柄保存到内存中

;BOOL WINAPI WriteConsole(
; _In_              HANDLE hConsoleOutput,
; _In_         const VOID  *lpBuffer,
; _In_              DWORD  nNumberOfCharsToWrite,
; _Out_             LPDWORD lpNumberOfCharsWritten,
; _Reserved_        LPVOID  lpReserved
;);
        sub   rsp, 8              ;对齐栈
        mov   rcx, qword[hFile]
        lea   rdx, [msg]          ;lpBuffer
        mov   r8, msglen          ;nNumberOfBytesToWrite
        lea   r9, [lpNumberOfBytesWritten]
        push  NULL                ;lpReserved
        sub   rsp, 32
        call  WriteConsoleA       ;如果成功,则返回非 0
        add   rsp,32+8

; BOOL WriteFile(
;       HANDLE       hFile,
;       LPCVOID      lpBuffer,
;       DWORD        nNumberOfBytesToWrite,
;       LPDWORD      lpNumberOfBytesWritten,
;       LPOVERLAPPED lpOverlapped
```

```
        ;);
        mov   rcx, qword[hFile]       ; 文件句柄
        lea   rdx, [msg]              ;lpBuffer
        mov   r8, msglen              ;nNumberOfBytesToWrite
        lea   r9, [lpNumberOfBytesWritten]
        push  NULL                    ;lpOverlapped
        sub   rsp,32
        call  WriteFile               ;如果成功则返回非 0
leave
ret
```

Windows API 文档使用成千上万个符号常量。这使代码更具可读性，并且更易于使用 Windows API，因此我们在程序开头包含文件 win32n.inc。这是所有符号常量及其值的列表。win32n.inc 文件可在以下位置找到：http://rs1.szif.hu/~tomcat/win32/。但是，请注意，在源代码中包含此文件将使可执行文件比需要的大得多。如果空间很重要，只需要在程序中包含所需的常量即可。如果使用 SASM，请找到安装 SASM 的文件夹，然后手动将文件复制到系统上的 SASM include 目录中。

我们在代码中将 Windows 函数调用的结构复制到注释中，以便跟踪正在发生的事情。我们根据调用约定将参数放入寄存器中，在堆栈上提供影子空间，调用函数，然后恢复堆栈指针。

如果一切顺利，函数 GetStdHandle 将返回一个句柄。否则，它将返回 INVALID_HANDLE_VALUE。为简单起见，我们不进行错误检查，但是在实际的生产程序中，建议你在程序中实施全面的错误检查。否则可能会导致程序崩溃，或者更糟的是，可能导致安全漏洞。

有了句柄之后，继续传递给 WriteConsoleA，传递句柄、要写入的字符串、字符串的长度、占位符(表示写入的字节数)和 NULL(表示保留参数)。前四个参数在寄存器中传递，第五个自变量被压入堆栈。这种压入将导致堆栈未对齐；将参数压入堆栈之前，我们必须预料到这一点。如果在压入之后对齐，则调用的函数将不会在堆栈上找到参数。在执行调用前，在堆栈上创建影子空间。

程序使用两种方法来写入控制台；一种是使用 WriteConsoleA，另一种是使用 WriteFile。WriteFile 使用相同的句柄，并将控制台视为另一个要写入的文件。在 WriteConsoleA 之后，为影子空间和对齐恢复堆栈。在 WriteFile 之后，不还原堆栈，因为这将由 leave 指令完成。

如果在 Windows API 文档中找不到 WriteConsoleA，请查找 WriteConsole。该文档指出有两个版本用于编写 ANSI 的 WriteConsoleA 和用于编写 Unicode 的 WriteConsoleW。

在 SASM 中运行此代码时，你将看到 WriteConsoleA 的第一个方法不起作用。函数在 rax 中返回 0，表示出现了问题。这是因为我们正在干扰 SASM 控制台本身。使用 WriteFile 的方法工作正常。

输出如图 40-1 所示。

```
Windows PowerShell
PS C:\Users\Jo\asm64win\02 helloc> make
C:\Users\Jo\asm64win\02 helloc>mingw32-make.exe
nasm -f win64  -g -F cv8 helloc.asm -l helloc.lst
gcc -o helloc.exe helloc.obj
PS C:\Users\Jo\asm64win\02 helloc> .\helloc.exe
Hello, World!!
Hello, World!!
PS C:\Users\Jo\asm64win\02 helloc>
```

图 40-1　helloc.asm 的输出

40.2　编译 Windows 程序

现在，将使用 Windows GUI，而不是使用控制台。将不提供完整的 Windows 程序，只是向你展示如何显示一个窗口。如果要执行更多操作，则必须深入了解 Windows API 文档。了解了它的工作原理后，只需要在 Windows API 文档中找到正确的函数，然后将参数传递到寄存器和堆栈中即可。

代码清单 40-2 展示了示例代码。

代码清单 40-2：hellow.asm

```
; hellow.asm
%include "win32n.inc"
extern ExitProcess
extern MessageBoxA

section .data
    msg db 'Welcome to Windows World!',0
    cap db "Windows 10 says:",0

section .text
    global main
main:
push rbp
mov rbp,rsp

; int MessageBoxA(
;     HWND hWnd, owner window
;     LPCSTR lpText, text to display
```

第 40 章 ■ 使用 Windows API

```
;           LPCSTR lpCaption, window caption
;           UINT uType window behaviour
;         )
        mov rcx,0              ; 没有窗口(window)所有者
        lea  rdx,[msg]         ; lpText
        lea  r8,[cap]          ; lpCaption
        mov r9d,MB_OK          ; 带有确定(OK)按钮的窗口
        sub  rsp,32            ; 影子空间
        call MessageBoxA       ; 如果选择了 OK 按钮，则返回 IDOK=1
        add rsp,32
leave
ret
```

输出如图 40-2 所示。

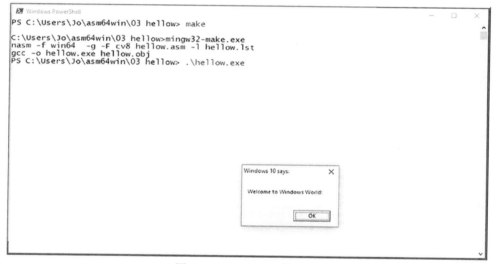

图 40-2　hellow.asm 的输出

当然，你可以质疑汇编语言是不是为 Windows 程序构建 GUI 的正确编程语言。使用 C 或 C++来实现计算密集型部件的调用更容易。

无论如何，你都应该读一本有关使用 C 或 C++进行 Windows 编程的好书，其中解释 Windows API，并通过提供正确的寄存器然后按所示调用函数将所有函数调用转换为汇编。需要诸如错误检查的复杂功能，而使用高级语言进行开发则要容易得多。

40.3　小结

本章内容：
- 如何使用 Windows API
- 如何将消息写入 Windows CLI(PowerShell)
- 如何使用指令 GetStdHandle、WriteConsole 和 WriteFile
- 如何创建一个带按钮的窗口

第 41 章

Windows 中的函数

当你有四个或更少的非浮点参数时，向函数传递参数会很简单。可以使用 rcx、rdx、r8 和 r9，并在调用函数之前在堆栈上提供影子空间。调用后，重新调整影子空间的堆栈，一切正常。如果有四个以上的参数，事情就会变得复杂。

41.1 使用四个以上的参数

下面首先看看为什么有四个以上的非浮点参数会使事情变得复杂，如代码清单 41-1 所示。

代码清单 41-1：arguments1.asm

```
; arguments1.asm
extern printf
section .data
    first    db "A",0
    second   db "B",0
    third    db "C",0
    fourth   db "D",0
    fifth    db "E",0
    sixth    db "F",0
    seventh  db "G",0
    eighth   db "H",0
    ninth    db "I",0
    tenth    db "J",0
    fmt      db "The string is: %s%s%s%s%s%s%s%s%s%s",10,0
section .bss
```

```
section .text
      global main
main:
push    rbp
mov     rbp,rsp
        sub     rsp,8
        mov     rcx, fmt
        mov     rdx, first
        mov     r8, second
        mov     r9, third
        push    tenth        ; 现在开始压入
        push    ninth        ; 相反顺序
        push    eighth
        push    seventh
        push    sixth
        push    fifth
        push    fourth
        sub     rsp,32       ; 影子空间
        call    printf
        add     rsp,32+8     ; 恢复栈
leave
ret
```

指令 sub rsp,8 之所以出现在这里,是因为当我们调用 printf 时,堆栈需要 16 字节对齐。为什么不在调用之前使用一条指令,例如 sub rsp,40? 堆栈将 16 字节对齐,但是 printf 可能会失败。如果我们在调用之前将堆栈减少 40 而不是 32,则堆栈上的参数不在 printf 期望的位置,而是在影子空间之上。因此,需要在开始压入参数之前对齐堆栈。请注意,我们需要以相反的顺序压入参数。调用之后,我们将为对齐和影子空间恢复堆栈。

输出如图 41-1 所示。

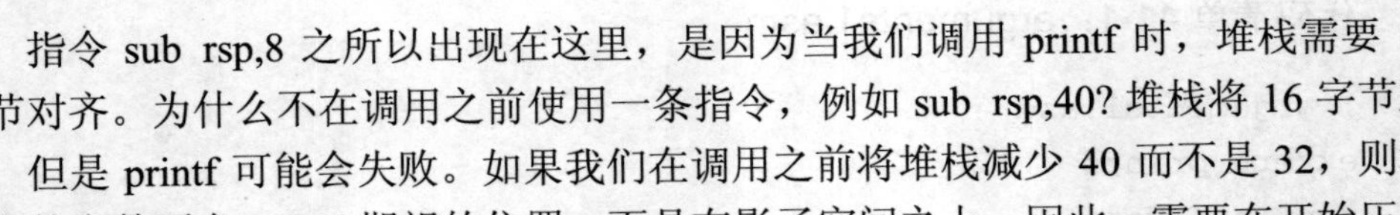

图 41-1　arguments1.asm 的输出

也可以用另一种方法构建堆栈。代码清单 41-2 展示了它的工作原理。

代码清单 41-2：arguments2.asm

```
;arguments2.asm
extern printf
section .data
        first     db "A",0
        second    db "B",0
        third     db "C",0
        fourth    db "D",0
        fifth     db "E",0
        sixth     db "F",0
        seventh   db "G",0
        eighth    db "H",0
        ninth     db "I",0
        tenth     db "J",0
        fmt       db "The string is: %s%s%s%s%s%s%s%s%s%s",10,0
section .bss
section .text
        global main
main:
push    rbp
mov     rbp,rsp
        sub     rsp,32+56+8      ;影子空间 + 堆栈上的 7 个参数 + 对齐
        mov     rcx, fmt
        mov     rdx, first
        mov     r8, second
        mov     r9, third
        mov     qword[rsp+32],fourth
        mov     qword[rsp+40],fifth
        mov     qword[rsp+48],sixth
        mov     qword[rsp+56],seventh
        mov     qword[rsp+64],eighth
        mov     qword[rsp+72],ninth
        mov     qword[rsp+80],tenth
        call    printf
        add     rsp, 32+56+8     ;在 leave 之前不需要
leave
ret
```

首先使用 sub rsp,32+56+8 调整堆栈。
- 影子空间为 32 个字节
- 压入的 7 个参数乘以 8 个字节，总共 56 个字节

然后，开始构建堆栈，并且当你看到必须对齐堆栈时，必须从堆栈指针中减去另外 8 个字节。

现在在堆栈的底部，有 32 个字节作为影子空间，在该空间的上面是第四个参数，再上面是第五个参数，以此类推。这里构建的堆栈看起来与上一个程序中的堆栈相同。你可以根据自己的喜好决定。

输出如图 41-2 所示。

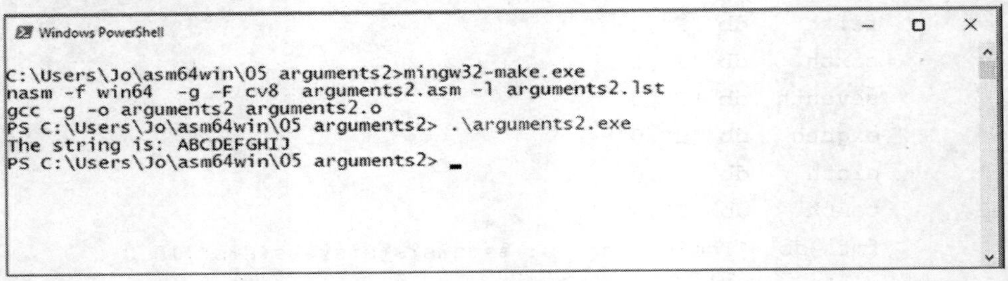

图 41-2　arguments2.asm 的输出

在被调用的函数中这是如何工作的呢？代码清单 41-3 展示了一些示例代码，这些示例代码使用 lfunc 函数来构建要由 printf 打印的字符串缓冲区。

代码清单 41-3：stack.asm

```
extern printf
section .data
        first    db "A"
        second   db "B"
        third    db "C"
        fourth   db "D"
        fifth    db "E"
        sixth    db "F"
        seventh  db "G"
        eighth   db "H"
        ninth    db "I"
        tenth    db "J"
        fmt  db  "The string is: %s",10,0
section .bss
        flist  resb 14         ;包含终止 0 的字符串长度
```

```
section .text
    global main
main:
push    rbp
mov     rbp,rsp
    sub rsp, 8
    mov rcx, flist
    mov rdx, first
    mov r8, second
    mov r9, third
    push tenth          ; 现在开始压入
    push ninth          ; 相反顺序
    push eighth
    push seventh
    push sixth
    push fifth
    push fourth
    sub rsp,32          ; 影子空间
    call lfunc
    add rsp,32+8
; 打印结果
    mov rcx, fmt
    mov rdx, flist
    sub rsp,32+8
    call printf
    add rsp,32+8
leave
ret
;-----------------------------------------------------------------
lfunc:
push    rbp
mov     rbp,rsp
    xor rax,rax                 ;清除 rax（尤其是高位）
    ;寄存器中的参数
    mov al,byte[rdx]            ; 把内容参数移到 al
    mov [rcx], al               ; 把 al 保存到内存
    mov al, byte[r8]
    mov [rcx+1], al
    mov al, byte[r9]
```

```
            mov [rcx+2], al
            ;栈上的参数
            xor rbx,rbx
            mov rax, qword [rbp+8+8+32]    ; rsp + rbp + 返回地址+ 影子空间
            mov bl,[rax]
            mov [rcx+3], bl
            mov rax, qword [rbp+48+8]
            mov bl,[rax]
            mov [rcx+4], bl
            mov rax, qword [rbp+48+16]
            mov bl,[rax]
            mov [rcx+5], bl
            mov rax, qword [rbp+48+24]
            mov bl,[rax]
            mov [rcx+6], bl
            mov rax, qword [rbp+48+32]
            mov bl,[rax]
            mov [rcx+7], bl
            mov rax, qword [rbp+48+40]
            mov bl,[rax]
            mov [rcx+8], bl
            mov rax, qword [rbp+48+48]
            mov bl,[rax]
            mov [rcx+9], bl
            mov bl,0              ; 终止 0
            mov [rcx+10], bl
    leave
    ret
```

main 函数与 arguments1.asm 中的相同；但调用的函数是 lfunc 而不是 printf，后者将在稍后的代码中调用。

在 lfunc 中，查看指令 mov rax, qword [rbp+8+8+32]，该指令将堆栈中的第四个参数加载到 rax 中。寄存器 rbp 包含堆栈指针的副本。堆栈上的第一个 8 字节值是我们在 lfunc 的序言中推送的 rbp。更高的 8 字节值是 main 的返回地址，当调用 lfunc 时它会自动压入堆栈。然后有 32 字节的影子空间。最后，我们得出推论。因此，第四个参数和其他参数可以在 rbp+48 和更高的位置找到。

当返回 main 时，堆栈再次对齐，并调用 printf。

图 41-3 展示了输出，当然与之前的结果相同。

```
PS C:\Users\Jo\asm64win\06 stack> make

C:\Users\Jo\asm64win\06 stack>mingw32-make.exe
nasm -f win64 -g -F cv8 stack.asm -l stack.lst
gcc -g -o stack stack.obj
PS C:\Users\Jo\asm64win\06 stack> .\stack.exe
The string is: ABCDEFGHIJ
PS C:\Users\Jo\asm64win\06 stack>
```

图 41-3　stack.asm 的输出

41.2　使用浮点数

浮点数是另一个故事。代码清单 41-4 展示了一些示例代码。

代码清单 41-4：stack_float.asm

```
; stack_float.asm
extern printf
section .data
      zero  dq 0.0    ;0x0000000000000000
      one   dq 1.0    ;0x3FF0000000000000
      two   dq 2.0    ;0x4000000000000000
      three dq 3.0    ;0x4008000000000000
      four  dq 4.0    ;0x4010000000000000
      five  dq 5.0    ;0x4014000000000000
      six   dq 6.0    ;0x4018000000000000
      seven dq 7.0    ;0x401C000000000000
      eight dq 8.0    ;0x4020000000000000
      nine  dq 9.0    ;0x4022000000000000
section .bss
section .text
      global main
main:
push rbp
mov rbp,rsp
      movq xmm0, [zero]
      movq xmm1, [one]
      movq xmm2, [two]
      movq xmm3, [three]

      movq xmm4, [nine]
```

```
        sub rsp, 8
        movq [rsp], xmm4

        movq xmm4, [eight]
        sub rsp, 8
        movq [rsp], xmm4

        movq xmm4, [seven]
        sub rsp, 8
        movq [rsp], xmm4

        movq xmm4, [six]
        sub rsp, 8
        movq [rsp], xmm4

        movq xmm4, [five]
        sub rsp, 8
        movq [rsp], xmm4

        movq xmm4, [four]
        sub rsp, 8
        movq [rsp], xmm4

        sub rsp,32  ; shadow
        call lfunc
        add rsp,32
leave
ret
;------------------------------------------------
lfunc:
push rbp
mov rbp,rsp
        movsd xmm4,[rbp+8+8+32]
        movsd xmm5,[rbp+8+8+32+8]
        movsd xmm6,[rbp+8+8+32+16]
        movsd xmm7,[rbp+8+8+32+24]
        movsd xmm8,[rbp+8+8+32+32]
        movsd xmm9,[rbp+8+8+32+40]
leave
ret
```

这个小程序没有输出，因为我们将在下一章中解释一个奇怪的地方。你将不得不使用调试器查看 xmm 寄存器。为方便起见，我们在注释中提供了十六进制的浮点值。前四个值传递给 xmm0 到 xmm3 寄存器中的函数。其余参数将存储在堆栈上。请记住，xmm 寄存器可以包含一个标量双精度值、两个打包双精度值或四个打包单精度值。在本例中，我们使用一个标量双精度值，为便于演示，我们将值存储在堆栈上，而不使用 push 指令。这就是将打包值存储在堆栈中的方法，每次以适当的数量调整 rsp 即可。更有效的方法是将标量值直接从内存压入堆栈中，如下所示：

```
push qword[nine]
```

在函数中，必须将堆栈中的值复制到 xmm 寄存器中，以便进一步处理它们。

41.3 小结

本章内容：
- 如何将参数传递给寄存器和堆栈中的函数
- 如何在堆栈上使用影子空间
- 如何访问堆栈上的参数
- 如何在堆栈上存储浮点值

第 42 章

可变参数函数

可变参数函数是一种参数数量可变的函数。printf 就是一个很好的例子。请记住，在 Linux 汇编中，当我们将 printf 与 xmm 寄存器一起使用时，约定是 rax 包含 printf 必须使用的 xmm 寄存器的数量。这个数字也可以从 printf 格式指令中获取，因此通常不需要使用 rax 就可以获得。例如，以下格式表示要打印四个浮点值，每个值有九个小数：

```
fmt db "%.f",9,"%.f",9, "%.f",9,"%.f",10,0
```

即使我们不遵守在 rax 中指定浮点值数量的约定，printf 仍将打印这四个值。

42.1 Windows 中的可变参数函数

在 Windows 中这个过程是不同的。如果前四个参数中有 xmm 寄存器，则必须将它们复制到相应的参数寄存器中。代码清单 42-1 展示了一个示例。

代码清单 42-1：variadic1.asm

```
; variadic1.asm
extern printf
section .data
      one     dq    1.0
      two     dq    2.0
      three   dq    3.0

      fmt     dq    "The values are: %.1f %.1f %.1f",10,0
section .bss
section .text
```

```
        global main
main:
push    rbp
mov     rbp,rsp
        sub     rsp,32              ; 影子空间
        mov     rcx, fmt
        movq    xmm0, [one]
        movq    rdx,xmm0
        movq    xmm1, [two]
        movq    r8,xmm1
        movq    xmm2, [three]
        movq    r9,xmm2
        call    printf
        add     rsp, 32             ; 在 leave 之前不需要
leave
ret
```

当在调用函数之前创建影子空间时，最好在执行函数后删除影子空间。在我们的示例中，add rsp,32 不是必需的，因为它紧跟在 leave 指令之前，该指令无论如何都会恢复堆栈指针。在本例中，我们只调用了一个函数(printf)，但是如果你在程序中调用了多个函数，请确保创建所需的影子空间，并且不要忘记每次继续使用函数后都删除影子空间。

在这里可以看到，将浮点值复制到 xmm 寄存器和参数通用寄存器中。这是 Windows 的要求，解释超出了本书的范围，但在使用非原型或可变 C 函数时是必需的。如果你注释掉了通用寄存器的副本，printf 将不会打印正确的值。

输出如图 42-1 所示。

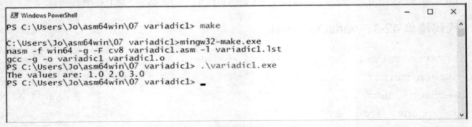

图 42-1　variadic1.asm 的输出

图 42-2 展示了不使用通用寄存器的输出。

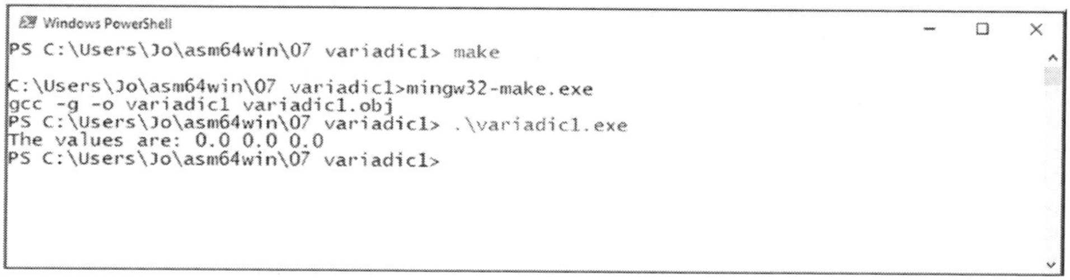

第 42 章 ■ 可变参数函数

图 42-2 variadiac1.asm 的错误输出

42.2 混合值

代码清单 42-2 展示了一个混合了浮点数和其他值的示例。

代码清单 42-2：variadic1.asm

```
; variadic2.asm
extern printf
section .data
        fmt     db "%.1f %s %.1f %s %.1f %s %.1f %s %.1f %s",10,0
        one     dq 1.0
        two     dq 2.0
        three   dq 3.0
        four    dq 4.0
        five    dq 5.0
        A       db "A",0
        B       db "B",0
        C       db "C",0
        D       db "D",0
        E       db "E",0

section .bss
section .text
        global main
main:
push    rbp
mov     rbp,rsp
        sub rsp,8               ;首先对齐栈
        mov     rcx,fmt         ;第一个参数
        movq xmm0,[one]         ;第二个参数
        movq rdx,xmm0
```

```
        mov   r8,A                    ;第三个参数
        movq  xmm1,[two]              ;第四个参数
        movq  r9,xmm1
; now push to the stack in reverse
        push  E                       ;第十一个参数

        push  qword[five]             ;第十个参数

        push  D                       ;第九个参数

        push  qword[four]             ;第八个参数

        push  C                       ;第七个参数

        push  qword[three]            ;第六个参数

        push  B                       ;第五个参数
; print
        sub   rsp,32
        call  printf
        add   rsp,32
        leave
        ret
```

如你所见，这只是一个遵守参数顺序的问题，在需要时将 xmm 寄存器复制到通用寄存器，并将其余参数以相反顺序压入堆栈中。

输出如图 42-3 所示。

```
Windows PowerShell                                                    —  □  ×
PS C:\Users\Jo\asm64win\07 variadic2> make

C:\Users\Jo\asm64win\07 variadic2>mingw32-make.exe
nasm -f win64 -g -F cv8 variadic2.asm -l variadic2.lst
gcc -g -o variadic2.exe variadic2.obj
PS C:\Users\Jo\asm64win\07 variadic2> .\variadic2.exe
1.0 A 2.0 B 3.0 C 4.0 D 5.0 E
PS C:\Users\Jo\asm64win\07 variadic2> _
```

图 42-3　variadiac2.asm 的输出

42.3 小结

本章内容：
- xmm 寄存器中前四个参数的浮点值需要复制到相应的通用寄存器。
- 如果有四个以上的浮点数或其他参数，则必须以相反顺序将它们存储在堆栈中。

第 43 章

Windows 文件

在 Linux 中，我们使用 syscall 来操作文件。在 Windows 中，我们必须遵循其他规则。如前几章所述，我们使用 Windows API。

代码清单 43-1 展示了示例代码。

代码清单 43-1：files.asm

```
%include "win32n.inc"
extern printf
extern CreateFileA
extern WriteFile
extern SetFilePointer
extern ReadFile
extern CloseHandle

section .data
    msg db 'Hello, Windows World!',0
    nNumberOfBytesToWrite equ $-msg
    filename db 'mytext.txt',0
    nNumberOfBytesToRead equ 30
    fmt  db "The result of reading the file: %s",10,0
section .bss
    fHandle                    resq 1
    lpNumberOfBytesWritten     resq 1
    lpNumberOfBytesRead        resq 1
    readbuffer                 resb 64
section .text
    global main
```

```
main:
push rbp
mov  rbp,rsp

;HANDLE CreateFileA(
; LPCSTR lpFileName,
; DWORD dwDesiredAccess,
; DWORD dwShareMode,
; LPSECURITY_ATTRIBUTES lpSecurityAttributes,
; DWORD dwCreationDisposition,
; DWORD dwFlagsAndAttributes,
; HANDLE hTemplateFile
;);
        sub     rsp,8
        lea     rcx,[filename]                      ;文件名称
        mov     rdx,GENERIC_READ|GENERIC_WRITE      ;所需的访问权限
        mov     r8,0                                ;没有共享
        mov     r9,0                                ;默认安全设置
                                                    ;以相反的顺序压入
        push    NULL                                ;没有模版
        push    FILE_ATTRIBUTE_NORMAL               ;标志和属性
        push    CREATE_ALWAYS                       ;配置计划
        sub     rsp,32                              ;影子空间
        call    CreateFileA
        add     rsp,32+8
        mov     [fHandle],rax

;BOOL WriteFile(
; HANDLE hFile,
; LPCVOID lpBuffer,
; DWORD nNumberOfBytesToWrite,
; LPDWORD lpNumberOfBytesWritten,
; LPOVERLAPPED lpOverlapped
;);
        mov     rcx,[fHandle]                       ;句柄
        lea     rdx,[msg]                           ;需要写入的 msg
        mov     r8,nNumberOfBytesToWrite            ;需要写入多少字节
        mov     r9,[lpNumberOfBytesWritten]         ;返回写入字节的数量
        push    NULL
        sub     rsp,32                              ;影子空间
```

```
        call    WriteFile
        add     rsp,32

;DWORD SetFilePointer(
; HANDLE hFile,
; LONG lDistanceToMove,
; PLONG lpDistanceToMoveHigh,
; DWORD dwMoveMethod
;);

        mov     rcx,[fHandle]           ;句柄
        mov     rdx, 7                  ;低 bit 位
        mov     r8,0                    ;没有高 bit 位
        mov     r9,FILE_BEGIN           ;从头开始
        call    SetFilePointer

;BOOL ReadFile(
; HANDLE hFile,
; LPCVOID lpBuffer,
; DWORD nNumberOfBytesToRead,
; LPDWORD lpNumberOfBytesRead,
; LPOVERLAPPED
;);

        sub     rsp,8                           ;对齐
        mov     rcx,[fHandle]                   ;句柄
        lea     rdx,[readbuffer]                ;要读入的缓冲区
        mov     r8,nNumberOfBytesToRead         ;需要读取的字节数量
        mov     r9,[lpNumberOfBytesRead]        ;已读取的字节数量
        push    NULL
        sub     rsp,32                          ;影子空间
        call    ReadFile
        add     rsp,32+8

                                        ;打印 ReadFile 的结果
        mov rcx, fmt
        mov rdx, readbuffer
        sub     rsp,32+8
        call printf
        add rsp,32+8

;BOOL WINAPI CloseHandle(
; _In_ HANDLE hObject
```

```
);
        mov rcx,[fHandle]
        sub rsp,32+8
        call CloseHandle
        add rsp,32+8
leave
ret
```

和以前一样，我们只使用 Windows API 函数的 C 模板来构建汇编调用。为了创建文件，只使用了访问和安全性的基本设置。文件创建成功后，CreateFileA 返回创建文件的句柄。请留意参数，你可以阅读 Microsoft 文档来了解不同的参数；有很多可能性可以帮助你微调文件管理。

文件句柄将在 WriteFile 中用于向文件写入一些文本。在第 40 章中，我们已经使用过 WriteFile 在控制台上显示一条消息。

将文本写入文件后，我们希望从位置 7 开始将文本读回内存，第一个字节的索引为 0。我们使用 SetFilePointer 将指针移到要开始读取的位置。如果 lpDistanceToMoveHigh 为 NULL，则 lDistanceToMove 是一个 32 位值，用来指定要移动的字节数。否则，lpDistanceToMoveHigh 和 lDistanceToMove 一起构成要移动的字节数的 64 位值。在 r9 中，我们指示从何处开始移动；可能是 FILE_BEGIN、FILE_CURRENT 和 FILE_END。

当指针设置为有效位置时，将使用 ReadFile 从该位置开始读取字节。所读取的字节存储在缓冲区中，然后打印出来。最后，我们关闭文件。检查工作目录，你将看到该文本文件已经创建。

输出如图 43-1 所示。

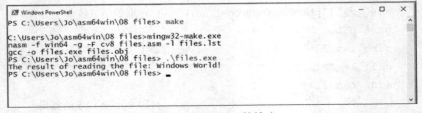

图 43-1　files.asm 的输出

43.1　小结

本章内容：

- Windows 文件操作
- 有很多参数可以帮助微调文件处理

后记

接下来做什么？

读完本书后，你已经掌握了现代汇编语言的基础知识。下一步取决于你的需要。本后记中提供了一些想法。

安全分析人员可以使用学习的知识来研究恶意软件、病毒和其他入侵计算机或网络的方式。二进制格式的恶意软件试图进入计算机和网络。你可以获取此二进制代码，对其进行逆向工程，然后尝试弄清楚该代码在做什么。当然，你需要在隔离的实验室系统中执行这样的操作。研究如何进行逆向工程并获取必要的工具。你应该考虑学习 ARM 汇编以分析智能手机上的代码。

作为高级语言程序员，你可以考虑构建自己的高速函数库以便与代码链接。研究如何优化代码；本书中的代码并不是为了高性能而编写的，而是出于举例目的。在本书中，我们提到了一些可以帮助你编写优化代码的概要信息。

如果你想全面了解英特尔处理器，请下载英特尔手册并进行研究。有很多有趣的信息需要理解和消化，知道硬件和软件如何协同工作将使你在开发系统软件或诊断系统崩溃方面具有优势。

作为一个掌握汇编语言的高级程序程序员，你现在可以更好地调试代码。分析你的 .obj 和 .lst 文件，并对代码进行逆向工程以查看会发生什么。查看编译器如何将代码转换为机器语言。也许使用其他指令更有效？